FLORIDA CITRUS 2004-2005 HURRICANES

John A. Attaway, Ph.D.

Published by
FLORIDA SCIENCE SOURCE, INC.
P.O. Box 8217
Longboat Key, FL 34228-8217

FLORIDA CITRUS 2004-2005 HURRICANES

Library of Congress Cataloguing-in-Publication Data

Attaway, John A., 1930-
Florida citrus 2004-2005 hurricanes / John A. Attaway
p. cm.
Includes bibliographical references and index.
ISBN 0-944961-09-6 (hardcover: alk. paper)
1. Hurricanes–Florida. 2. Citrus fruit industry–Florida. 3. Citrus–Effect of storms on–Florida. 4. Crops and climate–Florida. I. Title.
QC945.A76 2006
338.1'7430409759090511–dc22

2006024792

Current printing (last digit): 10 9 8 7 6 5 4 3 2 1
Published in the United States of America
E. O. Painter Printing Co., DeLeon Springs, Florida

FOREWORD

"Florida Citrus 2004/05 Hurricanes" is Dr. John Attaway's third book in a series on environmental issues facing Florida agriculture, and is the second book devoted to the impacts of hurricanes. This most recent book is a continuation of his previous literary project "Hurricanes and Florida Agriculture."

The hurricanes occurring in the first part of the 21st century have resulted in major economic damage to Florida's citrus industry. The author's efforts at documenting this damage after the 2004 hurricanes paint a stark picture of the potential havoc that can occur as a result of "Mother Nature's" wrath and their impacts on the citrus industry. In his previous book on hurricanes published in 1999, Attaway gathered a historical description of all recorded hurricanes. This book chronicles the detailed account of the ensuing hurricanes attacking Florida in 2004, namely Charley, Frances, Ivan, Jeanne and Wilma in 2005.

Valuable data and observations are presented on the meteorology, citrus production, marketing and aftermath of these devastating storms. Of special interest to the reader is the minute-by-minute account describing experiences of citrus growers who fought and lived through the ravages of hurricanes Charley, Frances, Ivan, Jeanne and Wilma. Pete Spyke, owner of Arapaho Citrus Management, vividly tells us how he and his family and employees survived the hurricanes that hit the eastern Florida counties and of the resourcefulness they exhibited before, during and after the hurricanes. Chet Townsend, owner of UltimateCitrus.com Citrus News, describes his impressions and assessments of the storms effects on citrus groves throughout much of the citrus growing area of the state. Both of their personal accounts and those of many others make for factual and interesting reading.

The author is a second generation Florida citrus grower who earned a Ph.D. degree in Chemistry from Duke University. He served as a scientist and administrator with the Florida Department of Citrus for 36 years as well as Adjunct Professor with the University of Florida's Citrus Research and Education Center at Lake Alfred.

An international authority on citrus, Attaway's extensive background and knowledge in science and agriculture serve as an important resource and insights in authoring this book. In addition to the meteorological aspects of hurricanes, he has addressed the effect of the 2004 and 2005 hurricanes on citrus production, economic value and on the orange juice futures market.

Every citrus grower, processor, packer, researcher, teacher, student of citrus and others concerned with hurricanes should read this interesting book.

Walter J. Kender, Ph.D.
Professor and former Director
Citrus Research & Education Center
University of Florida
Lake Alfred, Florida

On the Cover

Clockwise from top left: Hurricane Frances threatens the Florida east coast; Hurricanes Charley, Frances and Jeanne crisscross the state through Polk County; with 150 mph winds Charley did major damage to the Peace River Citrus Products plant; a DeSoto County orange tree after battering by Charley.

CONTENTS

NOTE FROM THE AUTHOR

The damaging effects of adverse weather on the Florida citrus industry have been a lifetime interest for me since I first torched an old-fashioned smudge pot in a family grapefruit grove north of Haines City on Sunday morning, November 26, 1950, when the temperature fell to 23°F. The freeze had been preceded by Hurricane King which roared up the peninsula October 17-18, 1950, lashing the grove with gale force winds. This had been an interesting year for weather, and had I not been committed to a chemistry major, I would have shifted to meteorology.

Forty-five years later and retired as Director of Scientific Research for the Florida Citrus Commission, a position I held from 1968-1995, I have had time to renew my interests in weather and have written two books. "A History of Florida Citrus Freezes," 368 pages, was published in 1997, and "Hurricanes and Florida Agriculture," 444 pages, was published in 1999. As I do my own research, and the first draft of each book was written by hand, I did not plan to undertake another lengthy manuscript. A friend had suggested a book on the damage caused by dry weather, but I vetoed that with the observation that who wants to read a book about drought. That's "like sitting and watching grass grow."

However, 2004, with three major hurricanes striking the heart of the Florida citrus industry in one year for the first time in history changed my outlook. For the sake of history, the destructive effects of these hurricanes should be recorded in the same manner as freezes in 1997 and historic hurricanes in 1999. I hope that this book will fill that need.

It had been my intent to end this book with Hurricane Jeanne and the collective effects of the three hurricanes, but when Hurricane Wilma struck Lee, Hendry, Collier, Martin, St. Lucie and adjacent counties, a number of people who had read my previous books asked that Wilma be included. Wilma was especially destructive as her full force struck Hendry and Collier counties which had been spared the onslaught of Hurricane Charley a year earlier.

DEDICATION

In Memory of Dr. James T. Griffiths who devoted 60 years to the Florida citrus industry and the problems of the small grower.

ACKNOWLEDGMENTS

Many individuals have been instrumental in the production of this book. Special gratitude is expressed to Mr. Peter Spyke of Arapaho Citrus Management, Fort Pierce, FL whose extensive contributions to Chapters 2 and 4, covering Hurricanes Frances and Jeanne, made it possible for readers to feel the daily problems of a citrus family before, during and after the hurricanes. Mr. Spyke is a great writer, and after he solves the problem of keeping St. Lucie County green we hope that he will write a book covering his lifetime experiences in the Florida citrus industry. Special gratitude also to Mr. Chet Townsend, publisher of UltimateCitrus.com Citrus News whose daily reports were a continuous source of up-to-date information, and to Mr. Kevin Bouffard for his articles in The Ledger and Ms. Susan Salisbury for her articles in the Palm Beach Post. Another major source of information was provided by the late Dr. Jim Griffiths through his newsletters written to members of Citrus Grower Associates and circulated to industry leaders.

Gratitude is expressed to Dr. Will Wardowski for his editorial expertise, to Mrs. Eileen Hoobin for her extraordinary word-processing skills without which the manuscript would not have made it to the printers, to Dr. Walter J. Kender for reviewing the manuscript, and to the E.O. Painter Printing Company who turned the manuscript into a book.

Special thanks to the many individuals who took time to sit down with me and relate their experiences from Hurricane Charley. From south to north, were growers Mr. Larry Hudson and Mr. Matt Sullivan in Charlotte County, Mr. Fran Becker of Peace River Citrus Products, Mr. Chris Bonnett of DeSoto Groves and Development and Arcadia grower Mr. Jeff Adams in DeSoto County, Mr. Keith Davis, Mr. Mark Manual, Dr. Stephen H. Futch, Multi-County Extension Agent, and Ms. Andrea Dunlap of the IFAS Range Cattle Station in Hardee County, Mr. Steve Callaham of Dundee Citrus Growers Association, Mr. Dennis Broadaway and Mr. Charles Counter of Haines City Citrus Growers Association, Mr. Steve Caruso of Florida's Natural Growers and Mr. Chris Oswalt, Citrus Extension Agent in Polk County.

For Hurricanes Frances, Jeanne and Wilma in Indian River and St. Lucie Counties, my thanks to Mr. Greg Nelson of DNE Sales, Mr. George Hamner, Jr. of Indian River Exchange Packers, and Mr. Scott Hurley and Mr. Jeff Caruso of Becker Holding Corporation,

For Hurricane Wilma in Palm Beach County, my thanks to Mr. Nat Roberts and Mr. Mark DuBois of Callery-Judge Groves, and for Wilma in Hendry and Collier counties, Mr. Mike Murphy, CEO of CPI, Coop Three and Ranch One.

At the Citrus Research and Education Center at Lake Alfred, Center Director, Dr. Harold Browning provided space where a retired scientist and research director could continue to write, Librarians Pam Russ and Marcia Alden who were always helpful, Gretchen Baut who helped with her experience in the photography lab and Sandy Hatch for his help in the Audio-visual area.

INTRODUCTION

The three hurricanes which devastated portions of the Florida citrus industry in August and September, 2004 were a unique phenomenon in Florida weather history. Whereas freezes had been a constant threat to our orange, grapefruit and specialty fruit crops, destructive hurricanes such as the 1949 hurricane which struck the east coast, and Donna in 1960 which marched up the peninsula from Ft. Myers to Flagler County, were relatively rare. Then from August 13 to September 24, three major hurricanes swept through the citrus belt in a period of only six weeks. As no two freezes are ever alike, so no two hurricanes are alike.

Each hurricane selected particular counties to ravage. Charley was most destructive to Charlotte, Lee, DeSoto and Hardee Counties, but also Polk County which had the unique distinction of feeling the eye of all three hurricanes. Polk County growers would have just as soon passed up the honor. Charley's winds were also damaging as he continued north-northeast through Osceola, Seminole, Orange and Volusia Counties.

Frances came from the east making landfall near Stuart, and did her greatest damage in Martin, St. Lucie and Indian River counties, but Okeechobee, Osceola, Highland and Polk counties also felt her wrath before she exited at New Port Richey in Pasco County. Frances was over peninsula Florida from her landfall east of Stuart at 0430 UTC on September 5 until she left the peninsula at New Port Richey early on September 6, a period of about 24 hours.

It is unusual that Jeanne also came ashore near Stuart in Martin County and followed almost in Frances' footsteps before turning north along the west coast through Hernando, Levy and Citrus counties. Jeanne made landfall at 0400 UTC on September 26 and entered the Gulf of Mexico at 1800 UTC on the same date. The greatest damage from Jeanne was to groves in Martin, St. Lucie, Okeechobee and Indian River Counties, but damage to groves in Highlands, Polk and adjacent counties was also significant.

In terms of ground speed, the three hurricanes also differed. Charley moved rapidly as it advanced up the peninsula and passed the author's home in Winter Haven in less than 4 hours. On the other hand, the tropical storm force winds of Frances blew over Winter Haven for 24 hours and Jeanne's winds lasted 8-9 hours.

Hurricane Wilma in 2005 was also unique as she followed a path from west to east across south Florida. Other hurricanes have followed a similar path, but were not common. Wilma came ashore near Marco Island in Collier County as a category 3 hurricane on the Saffir-Simpson Scale with winds of 125 mph, and exited between Stuart and Ft. Pierce on the border between Martin and St. Lucie County. It is interesting to observe that Wilma exited at what had been the entry points for Hurricanes Frances and Jeanne a year earlier.

An earlier hurricane to follow a similar path as Wilma was the hurricane of September 21-22, 1945 (Attaway, 1999, "Hurricanes and Florida Agriculture," pp. 182-188).

CHAPTER 1

HURRICANE CHARLEY, AUGUST 9-14, 2004

Hurricane Charley strengthened rapidly just before striking the southwestern coast of Florida as a Category 4 hurricane on the Saffir-Simpson Hurricane Scale. Charley was the strongest hurricane to hit the United States since Andrew in 1992 and, although small in size, it caused catastrophic wind damage in Charlotte County, Florida. Serious damage occurred well inland over the Florida peninsula.—Pasch, Richard J., Daniel P. Brown and Eric S. Blake, National Hurricane Center, 18 October 2004, revised 5 January 2005

The Meteorology

The system which ultimately developed into Hurricane Charley began as a tropical wave which emerged from the coast of Africa southwest of Dakar, Senegal on August 4, 2004, and is considered a "Cape Verde" hurricane although it did not become a named storm until it moved into the central Caribbean Sea early on August 10, 2004. The system had been slow to develop as it crossed the Atlantic, and had required 5 days before forming a tropical depression at 1200 UTC on August 9, about 100 miles south-southeast of Barbados.

Late on August 9 the depression moved into the southeastern Caribbean Sea and sped rapidly west-northwest at near 25 mph. Strengthening into Tropical Storm Charley as it moved west it continued to strengthen and became a hurricane with 75 mph winds as it approached Jamaica at 1800 UTC on August 11.

As a result, at 2100 UTC on the 11th of August a hurricane watch was issued for Florida Bay and Southwest Florida from Flamingo to Bonita Beach. At 0300 UTC on the 12th the hurricane watch was extended north of Bonita Beach to Anna Maria Island. At 0900, August 12th the hurricane watch was changed to a hurricane warning for the Florida Keys and the Dry Tortugas and southwest Florida from Cape Sable to Bonita Beach. A hurricane watch was issued from Bonita Beach north to Tarpon Springs. At 1500 UTC on the 12th a hurricane watch was issued from north of Tarpon Springs to the Suwannee River and at 2100 hours UTC the hurricane watch was changed to a hurricane warning from Bonita Beach north to Bayport. Later during the night the hurricane watch was changed to a hurricane warning from north of Bayport to the Suwannee River. Fearing the worst, many residents of the Tampa Bay area packed their bags, strapped their children into seatbelts and headed for what they considered would be the relative safety of Polk County and the Orlando area. Ironically, the Tampa area would be spared. They had moved from what would be a safe area directly into the final path of the hurricane up the Florida Peninsula.

Jamaica was lucky as the storm center passed 40 miles southwest of the island, and Charley shifted to a northwesterly course toward the Cayman Islands and western Cuba reaching Category 2 status on the Saffir-Simpson Scale near 1500 UTC on August 12. The Cayman Islands were spared the brunt of the hurricane as the center passed 17 miles to the northeast of Grand Cayman.

Turning to the north-northwest the center crossed western Cuba early on August 13 with maximum winds of 120 mph, emerging 15 miles west of Havana where it turned north over the Dry Tortugas at noon with maximum winds near 110 mph.

Leaving the Dry Tortugas behind, Charley then set a direct course toward Tampa, but an increase in southwest flow resulted in a change of course to the north-northeast in the direction of Charlotte Harbor. An increase in intensity to Category 3 on the Saffir-Simpson Scale followed with winds just over 110 mph. In the five hours prior to landfall, Charley's central pressure unexpectedly decreased from 28.49 inches to 27.79 inches of mercury bringing about a rapid intensification to the Category 4 level. With gusts near 150 mph at landfall across the barrier islands at 3:45 EDST, Charley was the strongest hurricane to hit the United States since Andrew in 1992.

Moving rapidly up Charlotte Harbor, Charley made a second landfall at Marjorie Point just southwest of Punta Gorda at 4:35 P.M. EDST and, by 5:30 P.M. EDST, had crossed Charlotte County and entered DeSoto County five miles west of Arcadia. Following the Peace River and U.S. Highway 17, the eye passed one mile west of Wauchula in Hardee County at 6:30 P.M. Holding a north-northeast course, Charley crossed U.S. Highway 17 between Bowling Green and Ft. Meade and passed four miles west of Lake Wales in Polk County at 7:30 P.M., where it spent almost an hour slashing through some of the most dense citrus groves along the central Florida ridge toward Haines City.

By 9:30 P.M. the storm center was passing through the greater Orlando area on a path to Daytona Beach where it exited into the Atlantic Ocean with winds still above hurricane force. Fast moving hurricane Charley had required only seven hours to pass from Charlotte Harbor to Daytona Beach following closely the path of devastating Hurricane Donna in 1960 (Attaway, 1999).

Charley's path west of U.S. 17 placed the town and cities from Arcadia north through Zolfo Springs, Wauchula, Bowling Green, Fort Meade and Lake Wales directly in the path of the devastating winds in the northeast quadrant. The storm cut a swath of winds 10 to 15 miles in diameter and exceeding 100 mph. Measured wind gusts included 127 mph at Punta Gorda in Charlotte County, 104 mph at Arcadia in DeSoto County and 109 mph at Wauchula in Hardee County, with an unofficial gust estimated at 136 mph reported near Ona in Hardee County. Florida's Natural processing plant at Lake Wales in Polk County measured a gust of 105 mph before losing its anemometer at 7:00 P.M. A spotter in Haines City reported a gust of 79 mph at 7:25 P.M.

The tiny eyewall of Charley had passed through many inland communities, towns and cities as it moved on its northeast course west of U.S. 17. These included Murdock, Pine Level, Lily, Limestone, Bridges, Oak Grove and Tomey. As the eye crossed U.S. 17 it made direct hits on Bowling Green, Ft. Meade, Alturas, West Lake Wales, Waverly and Dundee. The Historic Bok Sanctuary north of Lake Wales suffered major damage, indicative of tornados, but its beauty was fully restored by new trees and flowers by April 2005.

Moving on to the northeast, the storm passed well east of Haines City, Davenport and Loughman preventing major damage, but Campbell Station, Pine Castle and Winter Park suffered direct hits. Hurricane force winds were measured at 75 mph at Kissimmee in Osceola County and 105 mph at Orlando International Airport in Orange County.

Post Storm Data Acquisition (PSDA) Missions

Aerial, wind mapping missions were conducted by the National Weather Service (NWS) on August 15-16 and on August 29. Flights were made over the hurricane affected areas at 1,000 ft. and 2,000 ft. A NWS meteorologist was on the aircraft during the August 15-16 mission. A researcher from NOAA was added for the August 29 mission. Digital videos used subjective estimation based on Fujita Scale damage ratings to estimate wind speeds. Observations were limited in the area of highest winds due to lack of first hand knowledge regarding the construction of affected buildings as well as the characteristics of affected vegetation. For this reason, the results from this analysis are best regarded as estimates.

The PSDA missions showed gusts at landfall to be near 145 ± 15 mph, and that winds of 120 mph persisted inland to near Arcadia, after which 100 mph gusts persisted to the Orlando area, and 80 mph gusts to Daytona Beach where the hurricane exited the state.

Emergency management officials in Charlotte County reported extreme gusts at two locations in Punta Gorda, the Charlotte Regional Medical Center, 172 mph, and the Charlotte County Airport, 160 mph. There were no official National Hurricane Center (NHC) measurements. Damage measured with the Fujita Scale suggest winds at these locations to be in the 125-145 mph range.

The Rainfall

Since Hurricane Charley passed quickly after moving inland, it was not a heavy producer of rain. Most points near the hurricane center received about three inches of rain, with Kissimmee being the only location recording as much as five inches (Table 1-1).

Immediately after Hurricane Charley, inspection tours were made by many growers and interested parties including two well-known citrus industry veterans, Mr. Chet Townsend, Publisher of UltimateCitrus.com Citrus News, and Dr. James T. Griffiths, Managing Director of Citrus Grower Associates and publisher of the Citrus Grower Associates Newsletter.

Hurricane Charley to Hit Florida

As it became evident that Charley would hit the state, Chet Townsend in UltimateCitrus.com Citrus News, kept up a steady, daily stream of information for his subscribers beginning on August 12, 2004.

August 12, 2004. Most of the west coast of Florida is under a hurricane watch. At the time this newsletter was written local forecasters expected Hurricane Charley to be either a Catego-

TABLE 1-1. METEOROLOGICAL DATA FROM HURRICANE CHARLEY AS IT CROSSED PENINSULA FLORIDA FROM CHARLOTTE COUNTY TO VOLUSIA COUNTY ON AUGUST 13, 2004.

Location	Source	Peak Wind Gust (mph)	Rainfall (inches)
Naples	KAPF	55	1.75
Punta Gorda	KPGD	112	—
Fort Myers	KRSW	78	—
Ona	FAWN	—	3.09
Lake Alfred	FAWN	—	3.12
Kissimmee	KISM	75	5.20
Orlando	KMCO	105	2.11
Orlando	KORL	85	2.37
Sanford	KSFB	92	3.49
Umatilla	FAWN	—	2.88
Daytona Beach	KDAB	83	3.43

Source: Richard J. Pasch, Daniel P. Brown and Eric S. Blake. Selected surface observations for Hurricane Charley, pp. 1-8, Tropical Cyclone Report, Hurricane Charley, 18 Oct. 2004.

ry 2 (Winds 96-110 mph) or Category 3 (Winds 111-130 mph) storm by the time it makes landfall somewhere near Tampa and goes across north central Florida northeast to Jacksonville. If this track is maintained, most of the Florida citrus acreage should be spared a direct hit. However, a category 3 storm may be large enough to affect some citrus acreage in southwest Florida up through the Arcadia and Lakeland areas.

August 13, 2004. Most of the west coast of Florida is under a hurricane warning now. At the time this newsletter was written, local forecasters expected Hurricane Charley to be a Category 3 (Winds 111-130 mph) storm by the time it made landfall somewhere near Tampa and passed across north central Florida west of Jacksonville to the northeastern United States. If this track is maintained, the center of the storm should miss much of Florida's citrus acreage. However, a category 3 storm may be strong enough to affect some citrus acreage in southwest Florida up through the Arcadia (Peace River) area to Hillsborough and western Polk Counties. Local forecasters are predicting inland sustained winds of up to 60 mph. Rainfall amounts can be 4-8 inches on top of some already saturated soils along the west coast of Florida.

Hurricane Charley Damage Report

Saturday, August 14, 2004. Well, Hurricane Charley made landfall near Charlotte Harbor, between Punta Gorda and Port Charlotte with 145 mph winds on Friday afternoon. Damage there was extensive. Hurricane Charley then tracked across DeSoto County, near Arcadia, through Polk County, on to Orlando and then off the east coast near Daytona Beach. The Orlando Airport reported winds of around 85 mph with gusts estimated at 105 mph.

Citrus grove rainfall reports from www.agwise.net include 2.23 inches west of LaBelle, FL, 6.37 inches NW of Immokalee, and 5.30 inches in Devils Garden. AgWISE had very few rainfall reports from DeSoto County and the Ridge areas as the power went out and shut down the system. The only report from the Ridge was 1.79 inches at the CREC in Lake Alfred as of 6:30 P.M. On the east coast, all rainfall reports were less than 0.50 inches.

Very few damage reports on Florida citrus groves have come in yet. At my house near LaBelle, Florida with a small Valencia orange grove, I lost a few roof shingles and had some uprooted ornamental trees around the house and some large pine tree limbs on the ground. There are very few oranges on the ground at this time, but no uprooted citrus trees. I lucked out and also did not lose electrical power as was not the case in the Ft. Myers area where over 500,000 people lost their power.

At the time of writing this newsletter, I have only had one grower report: *From a grower southwest of Arcadia, Florida in DeSoto County*: "I surveyed my grove in Arcadia today. I just returned to the office. Although I had some uprooted trees and blown roofs, by far the biggest disappointment was to see the amount of fruit on the ground. Early estimate of fruit dropped is: ½ of Valencias and Fallglo, ⅓ to ½ of Murcotts, 80 to 90% of grapefruit."

As I mentioned, Hurricane Charley tracked right through DeSoto, Hardee, and Polk Counties with hurricane force winds. These counties are large orange producing counties and probably sustained significant loss of fruit in the path of the hurricane similar to the grower report above. The east coast (Indian River area) did not receive any significant damage. Southwest Florida (Hendry & Collier Counties) may have received spotty damage, but as a whole probably was spared of much major loss of fruit as the full force of the storm passed to the north of this large growing area.

Another Grower Report[1]

I took a trip over to Hendry, Charlotte, and DeSoto Counties today. I don't want to mention the specific groves I visited, because they're not mine. The closer to Arcadia, the more damage occurred. I didn't go north of Arcadia since I came home on SR 70. Basically, the citrus trees throughout the area survived the storm remarkably well. There were only a few mature trees tipped over—mostly on the ends of rows where they caught the full force of the wind. Most can probably be pulled back up. Very young resets were leaning, but these can be easily uprighted. Two-three year old trees that were growing vigorously were the most susceptible to being pushed over. They weren't snapped off, so they can probably be staked back up with very little permanent damage other than the labor costs involved.

I would say that the groves close to Arcadia lost quite a bit of fruit. The farther away from the path, the less fruit was lost on oranges and specialties. The fruit losses probably ranged from 10-30%. More will drop over the next couple of weeks, of course, but oranges seemed to come through fairly well.

The grapefruit losses were a different story. Even a good ways off the track, fruit losses were heavy. The 80-90% mentioned by the other grower is a reasonable estimate. I would say that there aren't many grapefruit left in the storm track. Speaking of the storm track, it was amazing how defined the track appeared to be. As I drove closer to the path, at one point there was a very sharp boundary between pines and oaks that were unaffected and those that were snapped off and stripped of leaves. It was like you drew a line—the transition was over maybe ¼ mile or so.

Arcadia looked rough. A lot of damage to every structure in town. About the only buildings that seemed to come through unscathed were the grand old heart pine houses in the old part of town. Power lines were laying everywhere. Power poles were snapped off like matchsticks. The amount of damage is almost inconceivable. I can't imagine what Punta Gorda and Port Charlotte must look like.

The East Coast is completely unscathed. Citra is fine, too. The town of Lake Wales took a big hit, which must have been a tornado because it wasn't that close to the storm track.

No doubt the fruit losses will be very variable, depending on whether the grove was in the storm or tornado path. Even within the groves, the fruit loss was variable—some rows didn't lose any fruit, while rows maybe 500 feet down the road might have lost 25-30%.

---

[1]Unidentified grower.

The grapefruit are the most haunting to me. It's unbelievable how much fruit loss there was. I have been through five or six hurricanes that have hit groves in my life, and I've never seen the fruit loss this great. Normally the fruit in the top ⅓ to ½ of the trees will get knocked off, but the trees provide a windbreak to their neighbors so the fruit nearer the ground comes through okay. Not the case with Charley—the fruit got knocked off all the way to the ground. It remains to be seen whether the fruit remaining on the trees has economic value—the picking cost may knock it out.

Citrus News Report on Hurricane Charley

Chet Townsend drove from LaBelle, in Hendry County, Florida, the second largest producer of oranges to Ocklawaha in Marion County, where very little citrus now exists since the freeze of 1983. He reported as follows in UltimateCitrus.Com Citrus News:

August 16, 2004. The LaBelle area received only minor damage from Charley. For the most part, the Hendry and Collier county citrus producing areas (the Gulf Citrus region) should not have any real significant losses of oranges. There may be some minor losses of grapefruit due to the strong winds, however. The East Coast (Indian River area) received no damage at all. The real damage to Florida citrus groves appears to be along the Peace River, up Highway 17 through Polk County. The eye of Hurricane Charley, made a right turn from the Gulf of Mexico into Charlotte Harbor, and traveled up along the Peace River Valley. It traveled towards the northeast across Florida south of Haines City, through parts of Orlando and out into the Atlantic Ocean near Daytona Beach. Where the eye passed through in DeSoto, Hardee and Polk counties, the damage seems to be severe with large fruit losses and major structural damage. (Polk is the largest orange producing county in Florida, and both DeSoto and Hardee are important orange producers. The three counties represent about 35% of Florida's orange crop.) Arcadia, Fort Meade, and Lake Wales were among the hardest hit inland cities.

Because the authorities were restricting access into Arcadia, I decided not to travel along Highway 17, but followed Highway 27 through Highlands and Polk counties northward. I was able to contact another grower near Arcadia who said citrus groves west and south of Arcadia were hit much harder than citrus groves east of Arcadia, which came through fairly well.

Along Highway 27 south of Lake Placid is where the first real signs of major wind damage showed up, mostly to billboards and older structures. When I stopped at an orange grove south of Lake Placid I saw no significant loss of fruit. In Lake Placid, there was a little damage to some signs, and the power was out. In fact, all of Highlands County was without power and was under a mandatory curfew from 9:00 P.M. through 7:00 A.M. Monday. All traffic lights were out from Lake Placid north to Interstate 4.

Beginning in Lake Placid and all the way north to Clermont, there were very long lines at a very few gas stations that had generator power. There was much debris along the road, with downed trees that had blocked Hwy 27, but had been cleared off to the side of the road. Many power lines were down, many poles were bent or snapped off.

Further north the worse the structural damage became. Sebring had a lot of damage to signs, billboards and a few structures. It was in Avon Park where I first saw some signs of fruit loss.

At a grove north of Avon Park, 10-15% of the fruit was on the ground, and this was probably within 15-20 miles from the track of the eye of the storm. New leaf growth was torn to shreds.

Overall, damage continued to increase on the way north. Along Highway 27 north, there was extensive damage to signs, power lines, and structures. Older mobile homes were completely torn apart, but just across the highway, there was a brand new mobile home dealer with virtually no damage to their models. New Florida codes for both regular homes and manufactured homes, that are much tougher since Hurricane Andrew, are working. There was 15-25% fruit loss from Frostproof to Lake Wales and Dundee.

Driving through Lakes Wales along 27 revealed something not seen in many years, the red Donald Duck water tower at Citrus World. The Florida's Natural logo had been blown off and was lying in the median of the highway. There was also extensive damage to structures and signs near Lake Wales. Orange losses were pretty heavy here as Lake Wales seemed to have been hit very hard. There was still no electric power and many large oak and pine trees down. There were a few orange trees uprooted along Highway 27, but they were mostly trees already in a weakened condition. There were many citrus trees uprooted closer to where the eye passed over.

Hopefully, the processing plants in DeSoto, Highlands, and Polk counties had good back up generators to prevent any loss of refrigeration for their stored juice. North of Haines City damage started subsiding. Interstate 4 marked the dividing line from traffic lights not working to lights on the north side that worked. Lake County seemed to have no damage.

There is no doubt that the Florida citrus industry experienced a significant loss of oranges due to Hurricane Charley. It is still too early to estimate the real loss. Fruit may continue to drop over the next couple of weeks due to bruising. If I had to make an estimate to the entire Florida orange crop loss, I would guess 8%. If you used the pre-Charley Louis Dreyfus Citrus estimate of 230 million boxes, that would reduce the volume for 2004-05 to 211.6 million boxes. Orange Juice Futures were up Friday.

Preliminary Estimates of Fruit Loss from Hurricane Charley

Several Southern Gardens representatives flew over DeSoto, Hardee, Polk and Highlands counties on Monday, August 16. Virgil Grimes made a ground check before flying over the area. He said that "there was a 35 mile wide path noticeable from the air that tracked right through some of the prime orange growing areas in the Peace River and Ridge growing areas." After his ground inspection, Virgil said he thought the loss would be around 10 to 15 million boxes. However, after flying over the area, he revised his estimate of orange losses to be around 25 million boxes.

The pre-Charley crop estimate from Louis Dreyfus Citrus was 230 million boxes. Elizabeth Steger's estimate was 221 million boxes net after a slight adjustment from Hurricane Charley. If there is a 25 million box loss, then Florida's orange crop estimate may be very close to 200 million boxes, or even less. Southern Garden's in-house estimate, made before the hurricane hit, was even lower, at 205 million boxes. If their estimate was correct, then the loss from Hurricane Charley could very well bring the Florida crop estimate well below 200 million boxes.

The remaining oranges still hanging on those trees outside the eye of the storm may size up enough to offset part of the loss, as most fruit is golf ball size now. If there are a lot of large oranges, it may lower the solids a little from fruit processed from this area next winter.

There are few grapefruit trees grown on the heavily damaged area, but these were largely stripped of their fruit.

According to Bob Terry of the U.S. Department of Agriculture, "Charley ripped some trees out of the ground and probably caused the most crop damage since Hurricane Donna in 1960. DeSoto, Polk and Hardee counties were hit hardest, he said. According to government figures, these three counties produced almost a third of the 203 million boxes of oranges the state sold last year."

Rory Dubin, a major grower in Arcadia with 1300 acres of orange groves, said in a telephone interview that "his personal crop loss is probably in the range of 40 percent to 50 percent, with 5 percent of his trees broken in half from the main trunk or uprooted."

And according to the Produce News on August 17, 2004, "as much as 50 percent of the Florida citrus crop in the 30-mile wide area swathed by Hurricane Charley August 13 may be off the market in the coming season.

The current golf-ball sized, unripened, green fruit is still months away from harvesting, and rumor suggests that half the crop may have been blown off trees completely. Fruit remaining on the trees may not be saleable to the fresh market because of the wind beating it received.

Crop damage in these areas is expected to be substantial," said Greg Nelson, president of DNE World Fruit Sales in Ft. Pierce, FL. "It is difficult to say what percentage of the Florida crop the damage and loss represent at this point. The good news is that groves on the eastern side of the state were not affected. This includes the Indian River area where much of the citrus destined for the fresh market is grown."

Hurricane Charley had forced the cancellation of the 13th Annual Florida Expo Trade Show and Seminar Program. The National Guard set up its base of operations at the Lee Civic Center in Fort Myers to provide assistance to people and areas in need following Hurricane Charley. The Civic Center is the location for the Expo, which is organized by LaBelle-based Gulf Citrus Growers Association (GCGA); the University of Florida/IFAS Southwest Florida Research and Education Center, Immokalee; and Citrus Industry Magazine, Bartow. However, the reception and dinner slated for Wednesday, August 25, was still scheduled and included a modified program to address the state of the citrus industry following the storm. The dinner featured U.S. Senator Bill Nelson and Congressman Allen Boyd. The two toured the Florida citrus and agriculture areas most affected by Hurricane Charley, then reported their findings and insights on federal relief efforts underway as a result of the devastation.

In addition, through the GCGA Scholarship Foundation, a hurricane relief effort to assist citrus growers, farm workers, and their families affected by Hurricane Charley will be launched at the Expo dinner," said Ron Hamel, Executive Vice President/General Manager. "This will be an excellent way for members of the industry to come together to help those most affected by the hurricane."

Modest Storm Losses for Florida Juice Plants

On August 18th it was projected by Reuters that storm losses for Florida juice plants would not be severe. Florida juice processors sustained modest damage from Hurricane Charley, with only one plant suffering badly from the storm that ravaged Florida's citrus belt last week, a senior industry official said Wednesday.

Industry officials, including Lisa Rath, Executive Vice President of the Florida Citrus Processors Association, said it was too early to give an accurate assessment of the damage inflicted by Charley on the upcoming Florida 2004/05 orange crop.

In a report released on Tuesday, the Florida Agricultural Statistics Service said the storm "passed directly through three major citrus producing counties on Friday, causing extensive damage and loss to trees and green fruit on either side of the hurricane's path." It added: "Assessments of loss cannot be made at this time. Rainfall amounts ranged up to four inches (10 cm) during and after the storm, with some areas receiving more. Growers are not able to enter some groves at this time because of blocked roads and standing water." Rath said the industry "won't have a handle on the losses for a few more weeks."

A Florida state agricultural official said the state may be able to come up with a preliminary calculation by early next week. But Rath added that droppage from the trees caused by the storm and the health of the citrus trees will take longer to gauge after the pounding dished out by the storm.

Observations by Dr. James T. (Jim) Griffiths

Dr. Griffiths wrote the following in his Citrus Grower Associates newsletter:

Before discussing what I know and understand about Charley, let me suggest that any of you who don't have a copy of Dr. John Attaway's "Hurricanes and Florida Agriculture" will probably find it quite interesting and worthwhile to get a copy.

Copies may be obtained from Florida Science Source, Box 8217, Longboat Key, Florida 34228-8217. Their fax number is 941-383-4680 and their E-mail address is fssource@aol.com. It will give you an opportunity to compare the hurricane of mid-October 1944, with Donna of September 9, 1960, and Charley of this week.

I've had a good look at Hardee County, but I haven't been to DeSoto County or south. Charley took almost the same path as Donna and the storm of 1944. It essentially zeroed in on US 17 and followed it from Charlotte Harbor to Daytona Beach. The storm was apparently moving at close to 25 mph, or slightly faster. There's a distance of almost 175 miles from Punta Gorda to Daytona, and the storm took a little more than 6 hours to get there, so it was traveling at something close to 25 miles an hour or more.

Sustained wind speeds at 5:00 near Punta Gorda were at 140 mph. By the time the storm reached the Polk County line, winds were estimated as down to 115 mph. They dropped further to approximately 90 mph in the Orlando neighborhood and were down to 85 mph by the time they reached Daytona. Wind gusts would be expected to be anywhere from a few, to 25 or 30 miles faster than the sustained speeds, but this clearly indicates that as the storm was moving from DeSoto into Hardee County there were sustained winds of close to 130 mph and certainly wind gusts to 150, or in that neighborhood.

Charley must have crossed Polk County at about 100 mph and it appears that the eye of the storm was large enough to have encompassed both eastern Winter Haven, and western Lake Wales. It entered Polk County near Ft. Meade and left Polk County northeast of Haines City.

The one place that I could make the best comparison was the yard of the house which I occupied on Lake Eloise during Donna. My son, Roger, occupies it now. He has more trees down and is probably in worse shape on the west side of Lake Eloise than I was after Donna. We had one big hickory tree down then, but he has several large cypress and oak trees torn up and down. He won't lose any that large, but some almost that size.

I want to talk about crop losses. I've certainly seen orange groves in the Wauchula area that have close to 90% of their fruit on the ground. The older groves, which had some old trees and then everything from resets up to 7 or 8 years old as a result of Tristeza, were generally hurt. Winds can whip around in those groves and get to all the trees.

However, west of town I saw a grove that was probably 20 or 25 years old that was hedge-rowed north and south. The trees were probably 16 to 18 feet tall. They had been topped and were hedged on both sides. I would guess that this grove would probably have no more than 10% fruit on the ground. It was a dramatic contrast caused by solid hedge rows that obviously had an effect on breaking up the wind.

In a general way, we have damage commensurate with fruit size. Grapefruit lost a higher percentage than Navel oranges, than early oranges, and the Valencias. Tangerine trees were commonly loaded and many of them were young and limber. Many of these were split apart, because of the heavy fruit crop on them. In terms of actual tree damage, they may have a higher percentage tree damage than most other varieties. I don't have any real guess on percent of trees damaged, but they were certainly far worse in the DeSoto-Hardee area than elsewhere.

I made a trip starting with Alturas, east of Bartow, and went across the Alturas Babson Park Cut-off and then north into Lake Wales on old 17. Of course, grapefruit is far worse than oranges everywhere, but there's not much grapefruit left in this area.

In the case of oranges, at first, I didn't see much worse than I saw up in north Polk County and in the area west of Alturas, but as one goes east and approaches Lake Wales, there are more and more oranges on the ground. As I crossed US 27 and went up State Road 17, I saw substantially more damage. I'm not going to estimate the volume, but certainly it's much worse there than I had seen previously in Polk County. This ties in with the fact that the strong side of the storm went through Lake Wales. There are far more shade trees and other trees down in that area.

I haven't made specific observations on Navels versus hamlins, but I'm told that Navels are generally worse. I suspect this ties in with the concept of the larger the fruit size, at the time of the storm, the greater the loss is likely to be.

There are more trees ripped up or pushed over east of Alturas and along the Ridge south of Lake Wales than I've seen elsewhere in Polk County.

In my own grove, which is four miles northeast of Haines City, I would say that the loss in Valencias is negligible. The same thing is true of pineapples. I have a few Robinson tangerine trees that were split, but they probably don't represent 1% of the total tree population.

In my grapefruit I would guess I've lost about 100 boxes per acre across the board, but I don't believe it's much worse than that. In a general way, this represents about 2 boxes per tree. Because of Tristeza and Blight, a lot of trees have been replanted over the years. The old trees were more inclined to have a full two boxes under them. The trees that were planted since 1960 are fair-sized trees, and have much denser foliage, but have less fruit on the ground. This generally coincides with what I've seen elsewhere.

If we make two groups of counties, the group with maximum damage would include Hardee, DeSoto, and Charlotte. I would add to that a group of counties in which there is some damage. These include: Polk, Highlands, Manatee and Sarasota. The three hardest hit counties contain 20% of the orange acreage in the state and less than 5% of the total grapefruit acreage. In 2003, they produced about 44 million boxes, or approximately 21.5% of the total crop. When you add Polk, Manatee, Sarasota and Highlands, you'll find counties which contain 48% of the total orange acreage, 14% of the total grapefruit acreage, and which produced 53% of the orange crop and 17% of the grapefruit crop in 2003.

The Citrus Commissioners estimated a crop of 226 million boxes. Elizabeth Steiger has an estimate of 223 million boxes. Dreyfus has an estimate of 230 million boxes, so that we were somewhere in the 225-226 million box orange crop for next year based on these estimates.

The USDA estimators have already counted limbs in much of the storm area. They now have an opportunity to complete their counts for the entire state, count again where they've already counted, and in those doubled counts they can determine something about tree damage, because they'll know how many limbs are lost and they'll have a good idea of the pieces of fruit that were taken off by the storm. They'll also have an opportunity to check fruit on the ground in the future to determine additional loss caused by the storm. This is good information. They need to get it out as soon as they can. The crop size is the important thing, but it would be nice to better understand relationships between wind velocities in places and extent of damage. We have a chance to make some measurements here. We shouldn't miss the opportunity.

In order to lose 20% of a 225 million box crop, you have to have a loss of 45 million boxes. If the three worst hit counties have their normal share of the crop, they would have to have 90% orange loss to provide a 45 million box loss from the total crop. On the other hand, if you add in the other four counties and they have collectively a 50% loss, then we'd get to that 20% loss statewide. That doesn't appear to be very likely, but to get to less than 200 million, we only have to lose about 30 million boxes and this could be arrived at with a 25% loss across the board. It takes just a 70% loss in Hardee, DeSoto and Charlotte to get to that 30 million boxes.

Thus, I'll make the kind of an estimate that I like to make. We'll probably have a crop of something in the neighborhood of 200 million boxes, plus or minus 25 million boxes. That seems quite safe! What will you guess?

Because only 18% of the total grapefruit is in these seven counties, the effect on grapefruit, which was much heavier on the individually affected groves, is only going to help the balance between supply and demand.

Florida Citrus Crop Damage Estimate at Nearly $150 Million

On August 13, five days after Hurricane Charley went ashore in Charlotte County, the U.S. Department of Agriculture's Farm Service Agency (FSA) posted preliminary figures representing an estimated crop loss impact of Hurricane Charley on Florida citrus growers. Officials estimated that the storm reduced the overall citrus harvest for the upcoming season by 20 percent of the total crop, which would equate to nearly $150 million in crop loss given crop price levels for the past two seasons.

This figure was for fruit only and did not represent tree loss and re-planting costs or damage costs associated with barns, equipment, processing and packing facilities, which will increase the overall economic loss the industry will face. In addition, growers expected more fruit to fall within the next few weeks, further adding to the overall fruit loss.

"This is a very preliminary figure that we fully expect to increase as additional damage is discovered and industry losses are calculated," said Andy LaVigne, Florida Citrus Mutual's Executive Vice President/CEO.

The Florida Citrus Mutual Response

On August 20, 2004, Andy LaVigne, Executive Vice-President of Florida Citrus Mutual wrote the following letter to the membership:

Dear Florida Citrus Mutual members,

August 13, 2004, will be a date that is permanently ingrained in the history of the Florida citrus industry. As you have seen, Hurricane Charley ripped through a major citrus-producing region of our state, decimating groves in its path and devastating the lives of all impacted residents. While the industry deals with this traumatic blow, it is important to remember that people's lives have been turned upside down, both personally and financially.

On August 14, Mutual immediately began working to do whatever necessary for the growers in the area. Staff has been busy working to answer questions; meet with local, state and federal officials; field media inquiries and send assistance in whatever forms necessary.

With this in mind, Mutual has organized a series of grower meetings designed to help answer your questions about disaster assistance and relief, put you in contact with people who can assist you, and hopefully make a difficult process a little easier.

Meetings will be held the week of August 23 in Arcadia, Wauchula and Lake Wales.

Mutual will update growers on issues of concern in the aftermath of Hurricane Charley including: Tree and Crop Insurance information, Post-Hurricane cleanup procedures and current damage and crop loss estimations.

Invited agents include: the United States Department of Agriculture, the Florida Department of Agriculture and Consumer Services, the University of Florida Institute of Food and Agricultural Sciences, Citrus Research and Education Center, the Southwest Florida Water Management District and South Florida Water Management District.

Mutual has already begun discussions with state and federal officials to request short-term and long-term disaster assistance, and to determine the types of assistance that can be provided.

Grower Information Meetings

Approximately 600 people attended three Post-Hurricane Charley Grower Information Meetings hosted by Mutual. Representatives from the Florida Department of Agriculture and Consumer Services (FDACS), the Southwest Florida Water Management District (SWFWMD), UF/IFAS, Tedder, James, Worden and Associates, the Agency for Workforce Innovation and the United States Department of Agriculture (USDA) were in attendance to address concerns and answer questions.

Citrus Canker. It is as important as it has ever been to ensure proper canker decontamination procedures are followed in the aftermath of the hurricane. One of the most efficient methods of spread of citrus canker is wind driven rain, and it is impossible at this stage to determine where spread may have occurred. Workers, cleanup crews, insurance adjusters and anyone else entering or leaving a citrus grove must decontaminate.

Pesticide Program. There is the possibility of some issues concerning farm storage facilities for pesticides. If you have concerns about the loss of these facilities, contact should be made with the Chief of the FDACS, Bureau of Compliance Monitoring.

Water Management. An Executive Order has been issued stating that all SWFWMD water permits within the Emergency Area that were scheduled to expire prior to December 31, 2004, were extended through December 31, 2004. The reporting of water usage has also been waived until the end of the 2004 calendar year. For more information on water management, go to the SWFWMD website at www.swfwmd.state.fl.us.

Tree Health. In the aftermath of Hurricane Charley, it has been noted that well maintained trees that were healthy before the storm weathered the winds better than weak trees and branches, and that deeply rooted trees suffered less damage than shallow rooted trees. It has also been determined that small, young trees were less damaged than larger, older trees. Trees that were recently hedged and topped also suffered less damage than tall, overgrown trees. The greatest threat to your trees following a hurricane like Charley is water. Trees can suffer from a lack of water and an abundance of water in the same grove. If you have standing water in your groves UF/IFAS recommends the following considerations: 1) Establish drainage; 2) Assess damage; 3) Reset, clear and re-establish irrigation; 4) Fertilizer practices; 5) Fruit handling; and, 6) Pest actions for weeds, insects and diseases.

Tax Issue. Documentation of physical damage, as well as the cost of clearing, repairing, replanting and rebuilding is extremely important. The following are recommended steps to help recover from loss: 1) Create clear and detailed records of all extraordinary costs and expenses resulting from the storm; 2) Maintain a detailed log of labor hours expended for clean-up and repairs; 3) Create new general ledger accounts in which to segregate and accumulate costs and expenses related to the storm; 4) Keep detailed production and harvest records; 5) Organize last year's and current year's pre-storm production, harvest, yield and return information; and, 6) Keep a file of newspaper and industry trade journal articles that describe the extent of crop damage. For more information on tax issues, please contact Tedder, James, Worden & Associates.

Workforce Innovation. The Florida Agency for Workforce Innovation (formerly the Florida Department of Labor) has received federal funding to temporarily employ displaced workers for cleanup in the aftermath of Hurricane Charley. This could serve as a benefit for workers whose jobs are temporarily lost and could provide a means for keeping them employed until their citrus industry jobs are again available.

Farm Service Agencies. If you have damage to your grove, fruit or tree loss, you should report your loss to your county FSA office. Many relief programs are available from the Federal government and administered through FSA. It is important that you notify them of damage to your grove so that your name will be in the system when relief funds become available. For a list of local FSA offices, visit the Mutual website at www.flcitrusmutal.com.

Risk Management Agency. If you have crop or tree insurance it is essential that you notify your insurance agent of your damage so an assessment can be made of your loss. Also remember that tree insurance also covers tree damage and loss for excess water. Even if you don't think you have enough damage to warrant a claim from the high hurricane winds, you should report any damage you have so that it can be added to any potential future damage from flooding. For more information, contact your crop insurance agent.

Further Information from Mutual

Mutual Applauds Bush's Relief Plan

On August 27, President George W. Bush announced his plan to seek $225 million in disaster assistance for Florida citrus growers who suffered damage due to Hurricane Charley.

"We greatly appreciate President Bush and the Administration's prompt attention to assist Florida citrus growers during this difficult time." said Andy LaVigne, Florida Citrus Mutual's Executive Vice President/CEO.

To date, preliminary estimates indicate that 20 percent of the state's citrus crop has been lost due to the storm. Hurricane Charley decimated about 115,000 acres of groves in several large, citrus-producing counties including DeSoto, Hardee and Polk.

"President Bush's announcement is a good start in the hurricane recovery process in order to ensure the short-term viability of our industry," LaVigne said. "Long-term, we must ensure the preservation of the current citrus tariff, and we look forward to continually working with the Administration and Congress on this important issue."

Florida Citrus Mutual will work closely with U.S. Secretary of Agriculture, Ann Veneman and her team to ensure that the distribution of assistance funds is implemented efficiently and quickly.

"Florida citrus growers applaud the efforts of President Bush, Secretary Veneman, Gov. Jeb Bush, Florida Ag Commissioner Charles Bronson and Congressman Adam Putnam for their continual support of Florida's signature crop." LaVigne said.

This expedited announcement came as a result of Mutual's immediate discussions with federal and state officials following the storm. Mutual will continue to work with state and federal officials regarding disaster assistance programs for Florida citrus growers and other industry sectors impacted by the hurricane.

Details regarding the distribution of the $225 million are not yet available. However, Mutual will notify growers as soon as information is provided.

Mutual would like to thank Valencia Harvesting, Joe L. Davis and Ben Hill Griffin, III for hosting the Post-Hurricane Charley Grower Information Meetings and Syngenta for sponsoring the lunches.

Bronson Seeks Declaration of Agricultural Disaster

On September 1, Florida Agriculture Commissioner Charles H. Bronson delivered a letter to Gov. Jeb Bush, requesting the Governor seek the declaration of an agricultural disaster in 13 counties as a result of damage from Hurricane Charley. The counties asked to be included were: Charlotte, DeSoto, Hardee, Highlands, Lake, Lee, Manatee, Orange, Osceola, Polk, Sarasota, Seminole and Volusia.

The following is Commissioner Bronson's letter to the Governor:

Dear Governor Bush:

The high winds, heavy rains and multiple tornadoes caused by Hurricane Charley, a Category 4 storm when it came ashore on August 13, 2004, exacted an enormous toll on Floridians. While we mourn those who lost their lives, we also empathize with those individuals who lost jobs, property and personal belongings.

For Florida's agricultural community, the path of the storm brought it directly through some of Florida's most productive agricultural land. In these areas, fruit, trees, potted plants, and structural debris litter the ground. Florida's nursery and citrus industries suffered the largest losses; however, shellfish, tropical fish and tropical fruit also have been significantly impacted.

In addition, many of the areas vegetable growers had already conducted the pre-plant activities (field preparation, fumigation, laying tarps, etc.) that are necessary prior to planting their crops. Much of this was washed away in the storm and will now have to be re-applied resulting in significant additional costs for the growers.

Nearly 1.2 million head of cattle were in the direct path of the storm. Thankfully, initial damage assessments do not indicate a significant livestock loss, however, these operations also sustained considerable damage to fences and structures.

Without some type of assistance, these producers face an uncertain future that jeopardizes their long-term viability.

The announcement last week by President Bush on the use of current U.S. Department of Agriculture (USDA) resources to assist growers for lost crops and trees as well as debris cleanup was very welcome. It is possible, however, that resources beyond those currently available may be needed to address all of the agricultural needs.

While I am aware that twenty-five (25) of Florida's counties have already received Presidential Declaration of Disaster, this declaration does not rely on actual damage assessments for each agricultural producer. Although damage assessments are still being compiled, it is clear that many of our counties meet the threshold for a USDA Secretarial Disaster Declaration. A Secretarial Disaster Declaration will ensure that disaster assistance payments from programs included in the Supplemental Appropriations Bill go to those most affected on the basis of their actual damage as it requires a more intense on-the-ground review of each producer's damage.

Preliminary estimates indicate that thirteen (13) counties would be eligible for a Secretarial Agricultural Disaster Declaration. For this reason, I would appreciate your contacting USDA to request a Declaration of Disaster for the following counties: Charlotte, DeSoto, Hardee, Highlands, Lake, Lee, Manatee, Orange, Osceola, Polk, Sarasota, Seminole, and Volusia.

Governor, thank you for all that you have done to ensure Floridians are getting the assistance they need during this very difficult time for our state and for your continued support of Florida's agricultural producers. Please do not hesitate to call on me if you need additional information.

Sincerely,
CHARLES H. BRONSON
COMMISSIONER OF AGRICULTURE

Personal Observations from Growers

These observations are from interviews with growers in the most affected counties in order of damage, beginning with Charlotte County, the first to feel the onslaught from Hurricane Charley.

Charlotte County

Larry Hudson is a third generation member of an agricultural family with 6,000 acres in Charlotte and DeSoto counties. Mr. Hudson is a member of the Peace River Valley Citrus Growers Association. The family ranch was started by his grandfather in 1924, and at one time was an 82,000 acre holding. The ranch is chiefly a cow/calf operation but had 300 acres of citrus, largely hamlins but with some Valencias. After Hurricane Charley they are totally out of the citrus business. Their fruit was a 100% loss and the root systems of the trees were too damaged to bring back. All pump systems, all sheds, and all manifolds for the irrigation system were destroyed. The most severe part of the hurricane passed in about 1-½ hours, but tropical storm force winds persisted much longer. The peak was between 12:00 and 1:00 P.M. and the eye was estimated to be 30 miles wide.

The Hudson homestead is 12 miles south of Arcadia, 10 miles from Punta Gorda, and only a mile from salt water. Charley was still gaining strength from the shallow, hot water when it came up the harbor, left the Peace River and hit the ranch and groves. The old homestead lost the main house, the wash house and the barns, but out of 1,200 head of cattle and horses, there was only one injury, to a horse with a little cut, like a paper cut, over one eye.

They are only eight miles from the interstate and are now in the path of development.

Matt Sullivan is General Manger for 3,373 acres of citrus in the Bermont Section of northeastern Charlotte County down State Highway 31 from Arcadia. He is a member of the Peace River Valley Citrus Growers Association. Mr. Sullivan indicates that the family groves are dedicated to juice oranges such as hamlins, pineapples, Valencias and the "Tropicana varieties." They are not in the fresh fruit market. Their fruit is marketed through Tropicana Products in Bradenton and Peace River Citrus Products in Arcadia. The Sullivans are an old line Frostproof family.

As the Sullivan groves were on the east side of Hurricane Charley damage was substantial with 10% of the trees a total loss. Fruit losses were substantial. The Sullivan office was located in downtown Punta Gorda and was a total loss and has been abandoned.

When the National Hurricane Center forecast indicated that Charley was turning to the northeast and Punta Gorda would take a direct hit with a storm surge which might reach the second floor of their home, the Sullivan family moved to a house in the grove. Damage to the house in the grove was minimal, and the family came through without injury. When they returned to the Punta Gorda house they saw two semi's lying sideways on I-75, and a hot tub in the back yard of their home. To this day, they do not know the source of the hot tub.

After the hurricane a tree infected with canker was found in the Sullivan grove and 10,000 trees were pushed and destroyed, reducing the groves by 1500 acres.

DeSoto County

Jeff Adams is located at Pine Level, just across the Peace River west of Arcadia. Jeff indicates that he lost all of his fruit and one of every five trees. He fertilized and cleaned up a month after Charley. The groves recovered sooner than he expected and he was looking for a good crop in 2005-2006.

Jeff is a member of the Peace River Valley Citrus Growers Association.

Chris Bonnet manages DeSoto Groves and Development, Inc. including 400 acres of family owned groves west of Arcadia on the highway to Sarasota. Prior to Hurricane Charley the operation included gift fruit and fresh juice operations. In discussing the hurricane, Chris stated:

> In DeSoto County the major damage was between Kings Highway to the west of my groves and Highway 31 east of Arcadia. I had trees twisted off a foot above the ground and blown away either to pile up against other trees or into the ditches around the groves. The main ditch was full of trees with no roots, see Figures 1-4.
>
> Trees 14 feet or more took the brunt of the wind. Every orange was on the ground. Four year old trees were blown over, but we were able to straighten them up, and fastened them to stakes. I estimate that they lost a year of growth. The polypipe irrigation was blown to the middle of the grove if no trees were in the way.
>
> The biggest problem we faced was cleanup. Trees which were not blown into the ditch along the end of the grove were pushed into the canopy of another tree. An equipment shed built with 8" × 8" posts was flat to the ground. We also lost our gift fruit packinghouse, and our fresh juice processing plant.
>
> Charley was not a rainmaker. He left no standing water. From September 2004 on through the next ten months we had the best weather ever.
>
> When Charley came up the Peace River to Arcadia, I am convinced that the hurricane fed off the warm water in the river which is relatively wide between Arcadia and the Gulf of Mexico. We lost power for 20 days, and telephone service for 14 days.

Mr. F. F. (Fran) Becker is Director of Fruit Procurement at The Peace River Citrus Products Plant located on Highway 72 just west of Arcadia. The eye of Hurricane Charley passed one mile to the west putting the plant in the northeast quadrant of the hurricane. Although the plant did not have an anemometer, Charlotte County Airport, 15 miles to the southwest, recorded a wind speed of 140 mph before the building collapsed. Mr. Becker reported:

The wind speed at the processing plant was 140 mph or higher, and multiple tornados also occurred. The winds southeast from them did more damage than the later winds out of the northwest which occurred after the eye of the hurricane passed. The evaporator tower had been constructed to sustain winds up to 160 mph. This prevented even more damage than that which occurred. However, the roofs of the feed mill, the main juice room, the five tank farms and the dry storage building were lost and had to be replaced, Figures 1: 5-8.

When the hurricane struck we had 9.6 million gallons of frozen concentrated orange juice and grapefruit juice in the tank farms. Due to a courageous effort by employees, who worked many long hours, and the help of other processing plants, we were able to save all but 2% of the juice. Much of it was moved offsite to our Bartow facility, and the remainder was stored for us by Tampa Juice, Duda, Southern Gardens, Berry, Dreyfus, All Temp and other processors.

With the roofs blown off all five of our tank farms and the daily August temperature in Arcadia holding at 90° or higher, keeping the juice in the tanks at less than 32°F was a monumental problem. Fortunately within 48 hours, we were able to obtain a 1,200 kilowatt power generator, mounted on a semi-trailer, and by using panels, blue tarps and six inches of spray foam insulation we were able to bring the temperature down to 32°F within four to five days.

Florida Power and Light needed a place to park their equipment and our lot right beside the highway was convenient, so we loaned space to them. They reciprocated by having power restored to us within a week. We have considered having generators on site permanently but the fixed cost and the cost of maintenance would be prohibitive.

I can't say enough about our plant and office personnel. Although they had suffered major damage to their own homes, the homes of two families were a total loss, everyone came to work regularly every day and no one complained. I can remember our Director of Sales standing out in the heat serving drinks to everyone for two weeks.

Charley had passed by so quickly that by 6:00 in the evening my wife and I drove from our home in Sarasota to Arcadia and looked at a blue sky and a bright sun. Our first son, William Frances Becker was born on August 24, eleven days after the hurricane.

Charley was a dry hurricane, but it rained every day afterward for three weeks.

In a sense we were lucky to receive the first hurricane of the season as we were able to line up contractors before the second and third hurricanes arrived. As a result we were ready to run fruit by November 1 and we actually started on November 16, just three months after Charley blew the walls and roofs from most of our buildings.

In past years when people evacuated from the coast to escape hurricanes, they would drive their RVs to Arcadia and park in a big lot near downtown. This year it was a disaster for them as the lot was full of upside down RVs."

James E. Selph is a County Extension Agent in DeSoto county with an office in Arcadia. He recalls "that Hurricane Charley got from the Gulf to Arcadia so quick and the winds constantly changed direction. It was a "whipping" wind rather than a constant blow and was packed with small tornados.

"Due to the overall concern over fences on the highway, and roaming cattle, the DeSoto County Extension Office became the repository for $114,000 worth of barbed wire, posts, stables, hay and feed. These materials were stored in the Extension building for three weeks until moved out by the five women who worked at the office."

Hardee County

Dr. Stephen H. Futch is County Extension Agent IV-Citrus for both DeSoto and Hardee Counties. His unofficial estimate of crop loss immediately after Charley was 35%.

Dr. Futch indicated that "the hardest hit area of DeSoto County, along Highway 72 west of Arcadia, showed severe damage to oak and pine trees but that the number of citrus trees uprooted was less than expected. Damage to trees on the perimeter of groves was much worse than damage to interior trees. Groves with resets showed more damage than groves which were solid set.

As the hurricane while severe, was relatively small, damage was less north into Hardee County than in DeSoto County, and dropped off markedly east or west of the storm center. Near the center crop losses ranged 50-80%. In the Orange-Co block on Highway 70 east of Arcadia, crop loss was estimated at only 5-15%. The crop loss in this grove was actually greater in Hurricane Jeanne than in Hurricane Charley."

The Edgar Davis family operates the Florida Fertilizer Company, Inc. in Wauchula and they are also Hardee County growers. The Davis Family are members of the Peace River Valley Citrus Growers Association. Keith Davis remembers:

Everything was tied down the morning of Friday, August 13, 2004. The Kash and Karry store ran everybody out by noon and closed the store by 2:30. The power was off at 4:30 P.M. so I kept up with the storm using a battery TV and a satellite dish which held out despite the hurricane force winds.

Figures 1: 1-4. Show damage to C. P. Bonnet-DeSoto groves near the Peace River at Arcadia. DeSoto groves were some of the first to feel the wrath of Hurricane Charley as he made landfall on August 13, 2004. Source: C. P. Bonnet, Arcadia, FL.

By 5:30-6:00 the wind was really blowing, with gusts from both east and west as the eye of the storm passed. A vicious gust moved my pickup truck a foot, although it had been parked near the house. Fortunately the house came through okay, but the fertilizer plant had major damage. The roof was totally destroyed at one building, and the old building lost the skylights and the tin from the roof. Roll-down doors were lost from another building, but nothing else, Figures 1: 11-12. At one point we heard a train coming, but as we have no railroad it must have been a tornado.

We lost power from ten to fourteen days and the telephones were out for two weeks. The devastated area stretched from 7 miles east of Wauchula, halfway to Avon Park, to 7 miles west of Wauchula. The worst was over in an hour and a half. It was a dry storm, very little rain.

Fruit losses were very heavy in Hardee County, with a Hamlin grove which normally produces 600 boxes to the acre, yielding only 200 boxes, Figures 1: 13-14. All sheds for irrigation pumps were gone.

Ornamental nurseries were severely damaged but to my knowledge, Hardee County ranchers did not lose a single cow although the IFAS Range Cattle Station west of Wauchula recorded 70-75 mph winds with higher gusts.

For the first six weeks after Charley, the Hardee County schools operated for only two weeks. As a result of damage to the stadium, the high school football team, the Hardee County Wildcats (A.K.A. the Hardee County "Blue Tarps") was not able to play a single game at home. Their homecoming game was played at Avon Park. In the spirit of the times, Avon Park did not charge for the use of their stadium.

Not only did Charley do structural damage in Hardee County, it was very wet from August 13, thru September 30. We received 32 inches of rain.

There was an interesting occurrence involving one of our company's employees who was busy cleaning up his yard after the storm when a stranger drove up in a pickup truck and offered him a generator. He replied that he couldn't afford a generator. The stranger replied that the generator was his to use as long as he needed it with no charge. Along with the generator he left two drop cords and two cans of gasoline. The man was from Tampa and had bought the generator thinking that the hurricane would hit Tampa, but since it did not he thought it would be put to better use in Wauchula. He gave our employee a piece of paper with his phone number in Tampa and told him to call when he no longer needed the generator and he would come and pick it up, which he did three weeks later. Our employee never knew the man's name. What a wonderful example of man's humanity to man.

There was also an interesting episode involving Phil Turner's grove in DeSoto County, Phil is a customer of ours.

A van trailer was parked next to a 20 ft. wide rim ditch around his grove. There was a 12 ft. deer fence across the ditch. When a 100+ mph gust (or a possible tornado) struck the grove, it picked up the trailer, blew it over the ditch and over the fence where it landed on its side. The fence was not damaged.

Mark Manual is a fruit buyer and grower in Bowling Green, Hardee County and a member of the Peace River Valley Citrus Growers Association. Mark related that with the hurricanes, 2004-05 was the worst year ever for harvesting labor due to the lack of housing. Many trailers and frame houses used by the seasonal pickers were mostly destroyed. When the pickers heard the situation they didn't come to Hardee County. No place to live and a man couldn't make a day's pay off two boxes of fruit to a tree. There was more money in hurricane cleanup than in picking fruit.

From my view as a fruit buyer, I estimated that Hardee County growers lost 35% of their fruit in Charley, another 20% in Frances and another 20% in Jeanne. There was very little problem in qualifying for disaster relief money at the Tier one level, 75%.

In my young grove east of Wauchula, we propped up the trees after Charley, only to have them blow down again in Jeanne.

Before the storms, I bought a crop of oranges in bulk for $8,000. The hurricanes blew all the fruit out of the top half of the trees. The bottom half still looked okay.

So I said to myself, maybe it will not be a total loss, but when I came back two months later to pick the grove, most of the bottom half of the fruit on the trees was missing. There was only an orange strip of fruit left at eye level across the entire grove. It was a brief mystery as to what had happened.

The farmer's cow herd next door had walked through the fences which had been knocked down by fallen pine and oak trees, and chowed down on all the crop they could reach leaving only the strip half way up the canopy. It cost me $8,000 as there was not enough left to pick, but I had to laugh at that thin orange strip all the way across a ten acre grove.

The Hardee County football wildcats became known as the Hardee County Blue Tarps or "Hurri-cats" after Charley.

After the hurricane, a lot of people gathered in Nicholson Hardware looking for building supplies, particularly tin for roofs. One-third generation Hardee County grower out from Wauchula asked another "Joe, how did you do?" The answer was, Well, there was a pine tree through the front window, the roof was mostly blown off the house, an oak tree crushed the garage, the roof was blown off all the barns, the dog ran off and never came back, but I was lucky, I still have my health and family.

Polk County

Charles Counter is Director of Field Operations for the Haines City Citrus Growers Association (HCCGA), a member cooperative of Florida Natural Growers located in Polk County. Counter observed that Hurricane Charley, with winds exceeding 100 mph was the strongest of the three hurricanes which passed through Polk County in August and September 2004. By the time Charley reached Polk County the width of its path

Figures 1: 5-10. Show destruction at the Peace River Citrus Products plant at Arcadia including damage to tank farm no. 1, the dry storage areas, the freezer area and roof damage to various buildings. Cooling towers are shown in the background. Source: F. F. (Fran) Becker, Peace River Citrus Products, Arcadia, FL.

Figures 1: 11-14. Illustrate tree damage in Hardee county and damage at the Florida Fertilizer Company, Inc. plant in Wauchula. Source: Keith Davis, Florida Fertilizer Company

Figures 1: 15-17. The effect of Hurricane Charley on a fresh fruit packinghouse. Source: Steve Callaham, Executive Vice President and CEO, Dundee Citrus Growers Association.

through the Haines City area was estimated at 12 miles; somewhat less than the width as it passed through Charlotte and DeSoto counties. Due to its high windspeed, and the fact that Charley was first, it was the most destructive of the three hurricanes, particularly to grapefruit and Ambersweet oranges which were the most vulnerable. Fewer Navel oranges were blown from the trees and even fewer Hamlin and Valencia. Fallglo and Sunburst tangerines lost fewer fruit than Honey tangerines, probably because Honeys could not be harvested until much later and were vulnerable to the continued droppage which followed the hurricanes.

The post-hurricane period was very interesting to observe as Polk County had not had a serious hurricane since Donna in 1960, 44 years, and many growers and caretakers had no experience. Among the things noted were the varying periods of drop. First there was the immediate drop of fruit during the hurricane, then fruit with split and damaged stems began to fall, followed by punctured fruit and fruit with windscar which developed sour rot over four to six weeks which led to further droppage.

Through the months following a hurricane we noted that when fruit held to the tree despite split stems, the fruit did not size properly resulting in an unusual number of small size fruit at the time of harvest.

Plugging of fruit was the worst we had seen in many seasons. At first we thought it was carelessness by the pickers, but when we picked fruit ourselves we found that it had been damaged around the calyx making it more susceptible to plugging.

As a result of leaf loss due to the high winds of the storms, groves tended to go through the winter with a light canopy followed in the spring by a heavy bloom which also resulted in undersize fruit, particularly Sunburst, Hamlins and Robinsons which we estimated to be undersize by 40%. Leaf loss also resulted in lower minor elements such as zinc and manganese in the spring and summer to follow.

Polk County was fortunate in that damage was largely confined to leaves and fruit, with very little loss of trees and limbs except where tornados occurred.

While Charley was the most damaging hurricane, each of the other two storms would have been very damaging had they been the first to occur.

The following is from an interview with **Mr. Steve Callaham**, Executive Vice President and CEO of the Dundee Citrus Growers Association (DCGA), the largest citrus packing cooperative in Polk County. "DCGA suffered major damage to their two packinghouses located in Dundee and Lake Hamilton."

Damage to both houses exceeded $1.2 million, mostly to roof structures in the receiving area, the harvesting shop and the coloring rooms, all of which were rendered unusable.

We began picking fruit between storms Frances and Jeanne so fruit was in-house but we had no power and damaged roofs so everything was exposed and being soaked. Fortunately, our workers came through and the fruit was salvaged with the help of portable generators.

The workers had problems of their own at home, but they realized how important the business was to all of us, so they pitched in to get the fruit moved successfully.

Our individual growers whose groves were in the immediate path of the hurricane, lost 80-90% of their grapefruit, 40-50% of oranges and 10-20% of tangerines, but fortunately groves which were a few miles from the hurricane center suffered little to no losses.

Like every one else we had a shortage of labor to pick the fruit. Many regular harvesters went elsewhere when they heard about the hurricane damage. With less fruit per tree they lost interest and as a result picking costs were much higher.

The following is an interview with **Mr. Steve Caruso**, Chief Executive Officer of Florida's Natural Growers in Lake Wales, Polk County.

From my home in Orlando, I followed Hurricane Charley play by play Friday night August 13 on television. It became obvious that the Florida's Natural plant north of Lake Wales would receive a direct hit.

Saturday morning after the hurricane passed, I left for the plant and found that my first problem was just to get out of Orlando, it was a nightmare. When I finally got on I-4, I made good time, but when I turned south on US 27 it was slow going again. Telephone poles were across the highway and all the traffic lights were out.

When I called the plant I learned that the people were out of the safe room and were evaluating the damage. Our anemometer had registered a gust of wind to 105 mph during Friday evening before it ceased to function. The eye of the hurricane had briefly passed over the plant and a period of calm had been experienced.

When I arrived at the plant I found the Florida's Natural sign was in the median of Highway 27. It was not restored until September 2005. The plant gate on Washington Avenue was gone, and there was major damage to the fruit bins. Any where that the wind could get under a structure like the fruit bins there was damage, but sealed buildings came through okay.

Fortunately, we had changed our building code to 110 mph in 1986, and as a result the new roofs were undamaged. Two older roofs which had not been upgraded were damaged.

Our aseptic tank farm had been built to 120 mph in 1998, and it came through the hurricane in good condition even though the eye of Charley came right over us.

We are proud of our dedicated employees, marketing, sales, engineering, etc., who came to work even though they had major problems at home. Our backup generators were operational within 24 hours and we were able to keep our inventory cold for both concentrated juice and juice not from concentrate, and were shipping within two days.

As Florida Power was out of electricity for a time, **we were able to export power to them**. Natural gas was cut off but in 24 hours we had our cogenerator up.

Chris Oswalt, Citrus Extension Agent for Polk and Hillsborough Counties writes a monthly newsletter, *Citrus Notes*. After Hurricane Charley Chris made the following suggestions:

There appears to be a wide variation in the amount of severity of tree damage associated with Hurricane Charley in Polk County. Trees can be roughly divided into groups based on the extent of damage. 1) Trees with little damage or a limb or two broken off, 2) trees with severe limb breakage, and 3) trees partly or completely blown over by wind and trees broken off at the trunk.

Trees with light to moderate limb breakage can be pruned and allowed to grow back naturally. The amount and extent of limb breakage will dictate whether or not pruning needs to be done. If you choose the little or no pruning method, remember that poorly pruned or non-pruning of damaged trees may lead to rot problems later on.

Trees with severe limb breakage need to be assessed on a tree-by-tree basis. Severely damaged trees may need to be replaced. Other trees should be pruned back with careful attention to proper pruning techniques. This pruning may be closely related to the same type of pruning done for trees after a freeze.

Trees that have been completely or partly blown over present a special situation. If the tree is completely blown over inspect the root system for possible causes as to why this tree may have been more susceptible to blow-over compared to surrounding trees. In some cases there may be other underlying problems with the citrus tree root system that have contributed to the loss of the tree. This inspection may provide information that you would not have otherwise seen.

Blown over trees have also occurred in groves with little damage and may have been caused by either a microburst or tornado associated with the hurricane. Trees should be inspected and a case-by-case damage evaluation of the affected trees should be done. Trees with obvious root damage may need to be replaced. Healthy trees could be righted and the root system replanted. Righted tree canopies should be pruned into balance with the loss of citrus roots. A careful analysis of the costs associated with righting vs. replacing the tree should be done in determining the best course of action. Irrigation and fertilization production practices used after a freeze would help to reestablish these trees. Trees with old freeze damage or heart rot were in many cases snapped off at the trunk. Obviously the remaining root system may need to be removed and a reset can be planted to replace the missing tree.

The new flush pushing out, as I write this article, is more than likely the last flush of growth trees will produce before this win-

Figures 1: 18-23. Damage to the Florida's Natural grower processing plant in Lake Wales was severe, but less so than at the Peace River Citrus Products plant in Arcadia. The maximum gust at the Lake Wales plant was 105 mph, about 40 mph less than gusts at the Arcadia plant. Source: Steve Caruso, Chief Executive Officer, Florida's Natural Growers.

ter. It would seem important for growers to encourage the growth and development of this flush before the onset of winter.

Lake County

John Jackson is County Extension Agent IV-Citrus. As Hurricane Charley had been over land for a considerable distance when it reached Jackson's counties he indicated that damage was minimal except for those areas in Osceola, Seminole and Volusia counties which were in the direct path of the hurricane center.

In Lake County there was very little tree damage, and trees which were damaged were likely to have had tristeza or root rot. There was some fruit on the ground but it was not significant. Rain was no issue in the sand hills of Lake County as the water was needed to restore lakes to their normal levels. Groves in Osceola and Brevard Counties did lose fruit, particularly grapefruit, but did not undergo major tree damage.

Actually the greatest damage from Hurricane Charley in Lake County was to the fern industry which suffered major damage to shade houses needed to protect ferns from the hot, summer sun. Over 1,000 acres of shade houses were lost.

In Seminole County fruit damage was here and there, but near Oviedo which felt the hurricane's eye wall, as much as 30-35% of the fruit was on the ground.

After his 37 years of experience in the citrus industry, Jackson has concluded that healthy citrus trees are surprisingly tolerant to hurricane damage, much more so than oak trees.

The Continuing Impact of Hurricane Charley

On August 21, 2005, it was announced by Dr. Dan Gunter, Executive Director of the Florida Department of Citrus that the loss of revenue from the decreased crop size would require a reduction in the 2004-05 advertising budget of at least $4 million. The budget had already been reduced to $61 million, $10 million less than the 2003-04 budget due to the failure of the Florida Citrus Commission to approve a tax increase.

Citrus was not the only loser. Jerry Mixon, Jr., Haines City area blueberry grower observed that the blowing sands during the hurricane had "stripped his bushes of leaves. It looked like grasshoppers had been through."

Others who suffered included cattle ranches with damage to fences and livestock in Arcadia and Wauchula, and the loss of 500-600 acres in the Palmetto-Ruskin area. (Kevin Bouffard, The Ledger)

Kevin Bouffard. The following appeared in the By-Line by Kevin Bouffard of the Lakeland Ledger. Bouffard covers Florida agriculture for the Ledger readers, particularly the citrus industry.

Charley Broke the Rules for Citrus

Hurricane Charley may have rewritten the book on growing citrus in central Florida.

"According to the book, green oranges don't blow off the tree—that's what the old-timers have always told me," said Scott Young, the manager of Peace Valley Enterprises, a citrus grower and grove caretaker in Alturas. "Somebody's got to rewrite that book because they blew off."

Richard A. "Dick" Fort, Jr., the general manager of Peace River Packing Co. in Fort Meade, read that same book.

"I was not expecting the orange damage we had, and I really was not expecting the damage to our tangerines," Fort said. "The tangerines are so small, I didn't expect any damage."

Ford and Larry Black, the production manager at Peace River Packing, estimated Charley blew off at least one-third of their oranges and tangerines and up to 90 percent of their grapefruit. The company owns 1,200 acres of citrus, all but 100 acres in the Fort Meade area of Polk County.

Young estimated Charley took off 30 percent of the fruit in his family's 400 grove acres in Polk County. Their losses will only grow worse during the next week, as citrus growers clear out groves strewn with uprooted trees, dead branches, dropped fruit and other hurricane debris.

"A lot of the fruit has been punctured or damaged, and in about a week a lot more fruit will drop off," said Ben Hill Griffin III of Frostproof, one of the state's largest growers with more than 15,000 acres across Florida.

Black and Fort pointed out those future losses as they took a Ledger reporter on a tour of their devastated groves.

Most of the Honey tangerines still hung from the tree in one grove, but upon closer inspection, the stem and the fruit had clearly begun to separate. Black showed how other green tangerines came off the tree with a gentle tug, much less than the force normally required to pick even ripe fruit.

"In this grove, when you come back here in a week, 70 to 80 percent of the tangerines will be on the ground," Black said.

Growers and other citrus industry officials told The Ledger it will take weeks to gauge even a rough approximation of the damage to the 2004-05 citrus crop.

"Probably nobody will ever truly know what the damage to citrus is because nobody knows how many boxes

TABLE 1-2. WATCH AND WARNING SUMMARY FOR HURRICANE CHARLEY—AUGUST 11-14, 2004

Date	Time (GMT)	Action	Location
8/11	2100	Hurricane watch issued	East of Craig Key to Ocean Reef including Florida Bay and SW Florida from Flamingo to Bonita Beach.
8/12	0300	Hurricane watch issued	North of Bonita Beach, Florida to Anna Maria Island.
8/12	0900	Hurricane watch changed to Hurricane warning	Florida Keys from the Dry Tortugas to the Seven Mile Bridge and SW Florida from East Cape Sable to Bonita Beach.
8/12	0900	Tropical Storm Warning Issued	Florida Keys from Seven Mile Bridge to Ocean Reef including Florida bay and on the mainland west of Ocean Reef to East Cape Sable.
8/12	0900	Hurricane Watch Issued	North of Bonita Beach to Tarpon Springs
8/12	1500	Hurricane Watch Issued	North of Tarpon Springs to Suwannee River Florida.
8/12	1500	Hurricane Watch changed to Hurricane Warning	Cuban provinces of Pinar Del Rio, La Habana, Ciudad de la Habana, Matanzas, and the Isle of Youth.
8/12	1500	Hurricane Warning Discontinued	Jamaica
8/12	2100	Hurricane Watch changed to Hurricane Warning	North of Bonita Beach FL to Bayport.
8/12	2100	Tropical Storm Watch Issued	Jupiter Inlet FL north to Altamaha Sound Georgia including Lake Okeechobee
8/12	2100	Hurricane Warning Discontinued	Cayman Islands
8/13	0300	Hurricane Watch changed to Hurricane Warning	North of Bayport to the Suwannee River
8/13	0300	Tropical Storm Watch changed to Tropical Storm warning	Lake Okeechobee
8/13	0300	Tropical Storm Watch Issued	South of Jupiter Inlet FL to Ocean Reef and North of Atlamaha Sound to South Santee River South Carolina.
8/13	0900	Hurricane Warning Issued	North of the Suwannee river to the Steinhatchee River FL
8/13	0900	Tropical Storm Watch Issued	North of South Santee River SC to Cape Fear North Carolina
8/13	1100	Hurricane Warning Discontinued	Cuban provinces of Pinar Del Rio, La Habana, Ciudad de la Habana, Matanzas, and the Isle of Youth
8/13	1500	Tropical Storm Watch Issued	North of Cape Fear NC to Oregon Inlet NC including Pamlico Sound
8/13	1500	Hurricane Watch Issued	Flagler Beach FL northward to the Savannah River near the GA/SC border
8/13	1500	Tropical Storm Watch changed to Tropical Storm Warning	South of Cocoa Beach FL to Jupiter Inlet FL
8/13	1800	Tropical Storm Warning changed to Hurricane Warning	Cocoa Beach Fl to Altamaha Sound GA
8/13	1800	Tropical Storm Watch changed to Hurricane Warning	North of Altamaha Sound GA to South Santee River SC
8/13	2100	Watches/Warnings Discontinued	South of Jupiter Inlet along the Florida E coast to Ocean Reef and for all of the FL Keys
8/14	0100	Watches/Warnings Discontinued	All of the Florida West Coast
8/14	0300	Tropical Storm Warning Discontinued	Lake Okeechobee, FL
8/14	0600	Tropical Storm Warning Discontinued	South of Cocoa Beach FL to Jupiter Inlet FL
8/14	0900	Hurricane Warning Discontinued	Altamaha Sound GA south to Cocoa Beach FL

Source: Richard J. Pasch, Daniel P. Brown and Eric S. Blake. 2004. National Hurricane Center, Miami, FL unidentified grower.

we started out with," said Dan Gunter, the executive director of the Florida Department of Citrus in Lakeland. "There are numbers being guessed all over the board."

Lakeland-based Florida Citrus Mutual, the state's largest growers' representative, estimated Charley will reduce the season's citrus harvest by 20 percent, according to a statement released Thursday. The Florida Department of Agriculture and Consumer Services had released an identical estimate earlier.

That 20 percent figure includes oranges, grapefruit, tangerines and other citrus varieties. It is a statewide estimate that includes major growing areas such as the Indian River region and Hendry County, which Charley did not touch.

No damage figures are available for Polk, Highlands, DeSoto and other counties in Charley's path, said Casey Pace, a Citrus Mutual spokesperson.

No firm figures on tree loss are yet available, but growers along the central Florida ridge, including Polk, reported a minimal number of uprooted trees. The sandy ridge soil allows citrus trees to develop deeper roots.

The damage in Hardee, DeSoto and other "flatwood" areas is expected to be substantial because the higher water tables there mean shallower root systems.

There is also no specific estimate regarding the damage solely to oranges, by far the state's largest citrus variety and the industry's economic engine.

The initial citrus crop forecast from the U.S. Department of Agriculture is not scheduled for release until October 12. USDA employees began counting fruit on trees across the state at the end of July, but they will begin again in a couple of weeks, Gunter said.

"An unofficial estimate from Dr. Elizabeth Steger, an Orlando citrus consultant, put this season's orange crop at 223 million boxes," said Dr. Tom Spreen, an agriculture economist with the University of Florida. Another unofficial count by Louis Dreyfus Citrus, Inc., a Winter Garden juice processor, put the orange crop at 231 million boxes.

Florida growers produced 242 million boxes of oranges in the 2003-04 season and nearly 50 million boxes of all other varieties. Annually, about 95 percent of the orange crop is processed for juice.

If the unofficial estimates prove accurate, even a 10 percent loss in the orange crop to Charley would significantly boost the farm price juice processors pay for oranges in 2004-05, Spreen said.

A 20 percent reduction, which would put this season's orange crop around 180 million boxes, the lowest in more than a decade, "would mean farm prices would get a whole lot better. A crop that low would fall 50 million boxes short of meeting the U.S. orange juice demand, Spreen said.

Young, the Alturas grower, disagreed with Spreen's analysis. A major loss in the orange crop would cause Florida processors to import more Brazilian orange juice, he said.

"I don't see any immediate change in price anytime soon." Young said. "We were below the break-even price before (in 2003-04), so anything can help. I'm hopeful, but not very."

Young, Spreen and other growers agreed that only groves not affected by Charley would likely benefit from any increased farm prices. Any imaginable price increase would not compensate for the loss of fruit blown off the trees.

"If you lost 10 to 20 percent of your fruit, you might be ahead of the game," Spreen said. "If you lost 50 percent, no way."

Ben Hill Griffin III estimated he lost 30-40 percent of his oranges in Polk, Highlands, Hardee, DeSoto and Charlotte counties, all in Charley's path. But he lost nothing from groves in other counties. "Obviously, we have sustained significant damage, but I think we'll come through it," he said.

The outlook for fresh citrus is murkier. Fruit that survives during the next few weeks will have a significant amount of wind scarring and other cosmetic defects to the peel, which reduce the chances of making it to the fresh market.

U.S. consumers generally will not buy scarred fruit, even if the interior quality remains unaffected. Even slightly scarred fruit fetches lower prices on the fresh market.

Most of his Fort Meade groves "appear to be a 100 percent loss for fresh use," said Black of Peace River Packing. "The customer doesn't care we had a hurricane.

Chapter 2

HURRICANE FRANCES, AUGUST 25-SEPTEMBER 8, 2004

Frances was a Cape Verde-type hurricane that reached a peak intensity of category 4 on the Saffir-Simpson Hurricane Scale. It affected the Bahamas as a category 3 hurricane and the Florida east coast as a category 2 hurricane.—John L. Beven II, National Hurricane Center, 17 December 2004

The Meteorology

Frances was the second Cape Verde hurricane to affect the Florida citrus industry in 2004. It was the sixth named storm, the fourth hurricane and the third major hurricane of the 2004 Atlantic hurricane season.

As is frequently the case, Frances developed from a tropical wave which moved off the west coast of Africa on August 21, 2004. The wave developed into a tropical depression at approximately 0000 hours UTC, August 25, 2004, 655 nautical miles west-southwest of the southern Cape Verde Islands.

Moving to the west on the south side of the Bermuda-Azores high, the depression intensified and attained tropical storm status as Frances at 1800 UTC on the 25th and reached hurricane status near 1800 hours on August 26, 2004. Hurricane Frances continued to intensify and reached a peak of over 130 mph on August 28th at latitude 18°N, longitude 53°W. After weakening slightly on the 29th, Frances reintensified on the 30th of August and attained Category 4 status on the Saffir-Sampson Scale on August 31st with 144 mph winds as she passed north of the Leeward and Virgin Islands.

Shifting to a west-northwest course on September 1-2, the center of the hurricane passed just north of the Turks and Caicos Islands and the southeastern Bahama Islands maintaining winds in excess of 140 mph. A hurricane watch was issued for the Florida east coast at 0300 UTC on September 2 followed by a hurricane warning at 1500 UTC. Affected by a westerly shear on September 3, Frances weakened to a Category 3 hurricane over the central Bahamas and finally to Category 2 with winds of 90-100 mph over the northwestern Bahamas on the 3rd and 4th of September.

Steering currents weakened as Frances reached the northwestern Bahamas on September 4 and drifted westward across the Gulf Stream at less than 8 mph. Intensifying slightly and then weakening again, Frances finally make landfall over Hutchinson Island, Florida near midnight on September 5 as a Category 2 hurricane. The highest measured storm surge was 5.89 ft. above mean sea level on the Florida east coast at the St. Lucie Lock. The National Weather Service at Melbourne estimated storm surges at 8 ft. near Vero Beach and 6 ft. at Cocoa Beach.

Frances continued to weaken as she moved northwest across Florida passing first over Frostproof, and then between Bartow and Ft. Meade in Polk County, across northern Hillsborough County and emerging into the Gulf of Mexico as a tropical storm near New Port Richey in Pasco County early on September 6, 2004. It had taken the hurricane almost 24 hours to cross the Florida Peninsula from east to west. While Hurricane Charley had been a compact storm which moved quickly across the state, Hurricane Frances was described as "big as the State of Texas."

Frances continued its northwest course through the northern gulf of Mexico and made a second landfall in the afternoon in the Florida "big bend" near Tallahassee as a tropical storm.

Damage to Citrus Groves

The major damage to the citrus industry from Hurricane Frances was to the grapefruit groves in the Indian River area, particularly Martin, St. Lucie, Indian River and Brevard counties. Because of the size of the massive storm center, which was 80 miles in diameter, it took twelve hours for the entire hurricane eye to come ashore. Groves in the Indian River area were battered relentlessly by hurricane force winds which extended 85 miles outward from the center.

As Frances moved inland, groves in Polk, Osceola, Seminole, Okeechobee and Highlands counties were also affected, but inland groves only received tropical storm force winds and fruit losses were much less than on the Indian River. However, precipitation was intensive, and Frances will be remembered historically for its extremely heavy rainfall, especially across north Florida. Wind speeds and rainfall for selected Florida locations are shown in Table 2-1. Note that totals over 15 inches of rain were recorded in High Springs, Chassahowitzka and Port Orange and in excess of ten inches in Gainesville, Ocala, Putnam Hall and the south side of Lakeland.

The lowest pressure recorded in Florida during the passage of Hurricane Frances was 962.1 mb at the United States Army Corps of Engineers (USACE) station at

TABLE 2-1. DATA FROM HURRICANE FRANCES AS SHE CROSSED PENINSULAR FLORIDA FROM MARTIN COUNTY TO PASCO COUNTY, SEPTEMBER 5-6, 2004.

Location	Sustained Winds, mph	Peak Gusts, mph	Rainfall, in.
Brooksville, Hernando County	38	54	—
Cape Canaveral, Brevard County	36	55	—
Chassahowitzka, Citrus County	—	—	15.81
Citra, Marion County	—	—	5.56
Daytona Beach, Volusia County	56	75	9.96
Ft. Pierce (FCMP Tower), St. Lucie County	81	108	—
Gainesville, Alachua County	47	66	10.85
Hastings, St. Johns County	42	—	8.35
High Springs, Alachua County	—	—	15.84
Jacksonville Int. Airport, Duval County	44	62	9.13
Lakeland (South), Polk County	—	—	12.01
Lake Tohopekaliga, Osceola County	55	82	—
Lake Wales Ridge, Polk County	—	50	—
Live Oak, Suwannee County	—	—	8.10
Martin County	—	108	—
Melbourne, Brevard County		75	7.95
Ocala, Marion County	—	50	10.81
Orlando Int. Airport, Orange County	54	69	6.50
Port St. Lucie (Texas Tech Tower), St. Lucie County	77	96	—
Port Orange, Volusia County	—	—	15.05
Putnam Hall, Putnam County	—	—	11.64
Sanford, Seminole County	51	70	4.40
St. Petersburg, Pinellas County	44	58	6.27
Tampa, Hillsborough County	45	54	5.74
Umatilla, Lake County	—	—	8.86
Vero Beach, Indian River County	68	84	—

Source: John L. Beven II, National Hurricane Center, Miami, FL 17 December 2004.

St. Lucie Lock at 0600 UTC on September 5, 2004. A storm chaser on southern Hutchinson Island reported an unofficial pressure reading of 959.0 mb at 0525 UTC on the same day.

A total of 101 tornadoes were reported from Frances, 23 of which were in Florida. Frances was directly responsible for seven deaths, five in Florida. An estimated total property damage of $9 billion made Frances the fourth most costly hurricane in U.S. history, at that time, behind Andrew (1992), Charley (2004) and Ivan (2004).

On September 6, 2004, at 9:05 A.M. Chet Townsend e-mailed to UltimateCitrus.com readers from his home in LaBelle the following:

Electricity was restored to my home just before dawn this morning. This time I was only without power 36 hours instead of 5 days after Hurricane Charley. Many growers on the east coast will probably not be as lucky as power crews will need to rebuild much of their infrastructure before restoring power to homes that were not severely damaged. There are reports of structural damage to sheds, barns, homes, and businesses in the Indian River area. Mobile homes again received the most damage. It is this type of housing in which many farm workers live on the east coast as in the other farming areas in Florida. It will still be some time before much of the damage can be fully accessed.

There are no citrus crop damage reports in yet. Today will be the first time that some growers will be allowed to drive back into the Indian River area if they had evacuated. Many roads are under water in places and there are a lot of downed trees. It may be Tuesday or even Wednesday before any official reports come out as today is Labor Day, a U.S. holiday. Today most growers are first accessing the damage to their own homes. One reported that wind blown rain forced water through closed windows and doors and saturated the inside of his home.

Hurricane Frances, downgraded to a tropical storm while slowly crossing the state of Florida, dumped a lot of rain over much of Florida. Rainfall reports on AgWISE, Agrilink's private growers' weather network, are very spotty due to possible damage to weather stations on Florida's east coast and power outages across Florida preventing the data from being uploaded to the Internet. Some of the rainfall reports available for Sunday, September 5, 2004, included 5.50 inches near Crewsville, FL (which received almost 2 inches the day before), 1.50 to 3.30 inches between Immokalee and LaBelle, FL in SW Florida. There are unofficial reports from the Indian River area of over 18 inches.

At 1:57 P.M. on the same date, Townsend reported again to UltimateCitrus.com subscribers as follows:

The first preliminary report has come in from west of Ft. Pierce. A grower driving along Sneed Road reported seeing "a lot of grapefruit on the ground." Some trees that can be seen from the road may only have 10-20 pieces of fruit left on the tree. There are a lot of leaves blown off the trees. However, there are hardly any citrus trees uprooted. (Mostly just a few weak trees here and there.) That is good news. If a grower did lose this crop, at lest the trees are there to grow another crop. There are also many barns and sheds with all or parts of their roofs missing.

Second hand information from Vero Beach says that there will be some grapefruit to harvest this season. It is not a total loss on the Indian River. There are no tree losses and some grapefruit trees still have fruit left on them.

It is still much too early to properly access the crop loss due to Hurricane Frances. These first reports are very preliminary and there are no fruit loss guesses or estimates yet. Fruit loss from wind can often be worse around the edges of a grove, especially where they are exposed, such as those that may be seen from a highway. Inside the middle of a grove, where other citrus trees may offer some protection, there may be much less fruit drop. There can be inside fruit and fruit on the leeward side of the trees that will stay on to be harvested this season. We have to wait until growers can get into their groves and make a full assessment, Right now, many groves are just too wet from the heavy rains to drive through.

Meanwhile, on September 9 John Weitzig, Director of the Florida Agricultural Statistics Service in Orlando noted that the first U.S.D.A. citrus crop forecast was still scheduled to be released on October 12, so there were no official figures on crop size available at that time. As it turned out, any figures would have been meaningless as Hurricane Jeanne was yet to come.

Estimated Fruit Losses from Hurricane Frances

On September 10, 2004, Florida Citrus Mutual released the following estimate to the media: "While assessments are still on going, preliminary estimates indicate that Florida citrus growers have sustained more than $200 million in damage from Hurricane Frances. This damage comes only weeks after an estimated $285 million in economic loss from Florida citrus growers due to Hurricane Charley."

Hurricane E-Mails from Peter Spyke

Peter Spyke, Arapaho Citrus Management, is a well known grower and grove caretaker in the Ft.Pierce area. He also owns The Orange Shop on U.S. 301 north of Ocala near Orange Lake. Mr. Spyke has been active in citrus industry organizations for many years including serving as Chairman of the Florida Citrus Commission's Scientific Research Advisory Council. The following are edited portions from e-mails he wrote to friends, family and clients during Hurricane Frances. He speaks of his wife Cindy and employees of Arapaho Citrus Management and The Orange Shop.

Sunday, September 5, 2004. It's Sunday morning, and Cindy and I are down at my Mom's in Ft. Lauderdale, where we came to ride out the storm. We've managed to talk to a few people in Ft. Pierce who stayed there and rode it out. Here's what we know right now.

Our house that we use for an office in the grove on Okeechobee Road made it through OK, so that's where we'll be headed tomorrow after the storm finishes passing. We lost a pole barn in the home grove that collapsed on the hunting jeep. Some tin was peeled off of the shop, but it's still standing, which is a good thing because that's where the generator and chain saw are stored that we need to power the house and clear things out. The dog kennel made it through, too, so the dogs will have a home—they're with us down here right now.

My sister Barbara reports loss of part of the roof of the packinghouse down here, and things are a mess inside. Mom's house is fine, so we're in a good spot.

The Orange Shop (in Citra), at last report, was fine. Emory has rounded up about 200 gallons of gasoline, and will load up all the ice he can find along with some visqueen and tarps, and will head down with the fruit truck as soon as the roads are clear so that we'll have gas for our trucks. This will allow us to keep things moving in the groves with water removal, initial clean-up, and security. We have plenty of diesel for the tractors and pumps, so we'll be OK until the fuel companies can dig out and make some deliveries.

There is a 24 hour curfew in effect for all of St. Lucie County, so we will stay put here at Mom's until tomorrow. We lost power yesterday at her house, but got it back a little while ago so we will be able to take a hot shower, have a cup of coffee, and wash the wet clothes before heading back and living on generator power for some time.

Danny Yates, one of Arapaho's stalwart employees, rode out the storm in the groves, running pumps and fueling the tractors we use to run the pumps after the power went out at about 8:00 P.M. on Saturday. Apparently, about 50% of the fruit—a very rough estimate at this point—was lost at Shadow Lakes. The loss was probably heavier in Ft. Pierce, about 20 miles north of Shadow Lakes, since the north eyewall went through there, and hammered the groves with strong north, east, and then southeast winds at full strength. The trees on the north ends of the tree rows were damaged, but the trees in the rest of the rows were OK with only leaf and fruit loss. The leaves will grow back quickly, so the main issue at this point in the groves is determining the amount of fruit loss and starting insurance proceedings. Our planting layouts are designed to reduce storm damage, and it looks like it worked again since the losses appear to be less than expected.

We lost some of the tin roofs over the pumphouses at Shadow Lakes, but the main buildings made it through OK. The canker wash, built from 6-inch I beams, is twisted up like a pretzel.

We are very thankful. Everyone is OK, and there will still be something left to work with. Given the path and intensity of this storm, that's more than we hoped for.

Tuesday, September 7, 2004. Cindy and I drove the two 4-Runners up from Mom's in Ft. Lauderdale yesterday, and with traffic plus stopping to hunt gas we finally got to the Arapaho office at the grove near Ft. Pierce about 6:00 pm. As Danny Yates had reported, the house looked OK. There had been one minor leak in the back office, but otherwise things were intact.

The yard, though, didn't look so good. Our big pole barn in back had blown down, smashing everything it landed on. The only thing that appeared to survive was the johnboat I use for duck hunting. It made it because my Jeep pickup on one side and an old Bronco on the other broke the fall of the building.

The roof is gone on the dog kennels, and a bunch of tin was peeled off the shop. Everything was wet—the whole grove is flooded, as are most of the groves in the area, which is an old citrus area developed before modern pumping systems, etc. All of the newer groves are upstream, which also means uphill, so when their pumps are fired up it floods us out. Fortunately, the land is excellent, which helps the trees survive inundation a little better, so if it's gone in 3-4 days we should be OK.

The fruit crop is pretty well gone in the home grove, especially grapefruit and Navels, our two most popular varieties. Some of the juice oranges and tangerines hung on a bit better. I have not tried to work up an estimate yet because we have been focusing on people issues.

We got up early on Wednesday and began next phase of the Arapaho response for our customers' groves. We headed a backhoe down to Shadow Lakes to clear roads on the way so people could get through and start helping the water removal process at the grove. The rainfall from the storm had exceeded the design capabilities of the drainage system, so everything is running at absolute limits, and the tolerances are tight as bowstrings. So, the backhoe's job is to go around and find limitations in the drainage canals, and make them go away. The canals are full of debris, which traps floating aquatic vegetation and forms dams that restrict water flow. As each one is removed, there's another upstream so that's the next stop. We'll do that basically for the next 3 days, for as long as the operator can stand it. Tomorrow we'll start a second backhoe out in Ft. Pierce, with the same mission.

Most of the Arapaho crew have been dealing with problems. Robert's mobile home where he lives lost the roof, and then a tree fell in the living room. Nobody has power, and many don't have water. The stores don't have power either, and the people who work in the stores don't have power, and the gas lines at the few stations that are operating with generators are 2 hours long. FEMA brought in ice and meals ready to eat for 2,000 people, and 5,000 people showed up. The power crews coming in from out of state are caught in traffic jams on the interstates caused by all the people who are trying to get back home after evacuating.

The Nextels only work part of the time. The towers are damaged, so you keep getting cut off in the middle of conversations. Miraculously, the phone lines still work at the grove, so we have one old analog phone plugged in to the main office line so we can get calls if we're here. If not, and the generator has run out of gas, the answering machine won't work because it needs to be plugged in to power. Growers are calling like crazy, looking for help to stand up young trees that have blown over and unplug water furrow drains. Our hand crew is short because some of the guys took off and went across state to run chain saws cleaning up after Charley at double what we pay them. So, we're going to try and gather up everybody at the shop tomorrow, and make a plan. First priority is to get everyone situated with their homes, etc., and then we need to mobilize to deal with our customer's groves, and help out other growers to the extent possible.

Emory is going to try and find some ice again up in Citra. He's already bought 250 gallons of gas, and hopefully he can get down by the end of the day tomorrow so that we can send everyone home with a full tank of gas and enough ice to take care of their families. Those two things are the currency du jour right now.

Tomorrow, I'll get around and make an initial assessment of fruit losses and tree damage. I've already called the insurance companies to report claims. We'll need to document a lot of things, and come up with some bottom line statement for the reporters. Normalcy will not be anywhere near the citrus industry this year. Between Charley and Frances, the rules have dramatically changed. The choices that we'll have to make are not ones with which we have any experience. The season is only about 30 days away. We have to make some quick decisions.

Sunday, September 12, 2004. Wednesday morning I headed down to Shadow Lakes to check the grove. Most of the grapefruit and Navels were blown off. The lemon trees were actually broken up pretty badly since the fruit didn't blow off and the limbs had so much weight at the end. After seeing the lemons, it appears that losing the fruit on the grapefruit probably saved the trees. The rest of the varieties didn't lose too much fruit. In my personal groves, though, I have mostly grapefruit, Navels, Honeybells, and lemons. Ft. Pierce was hit harder than Stuart, so I will pretty well have a zero season this year. Shadow Lakes should do better, because there are other varieties and the grove itself wasn't hit as hard. There will be losses, though, so it will require some readjustment until next year at this time.

We worked on the generator some more Wednesday, and fashioned a remote gas tank that will now allow us to run about 2½ hours between refills. This allows us to relax even more, and most importantly, keep the exhaust fan running longer at night. The grove house was built in stages from the 20's through about 1956. All of it was well before air conditioning, so the house was designed to be as comfortable as possible without AC. One of the most important features is a large fan in the ceiling of the utility room that draws air from inside the house and exhausts it into the attic. This does two things. First, the air inside the house is kept moving, and if you manage the open windows and doors right, there will be a breeze where you are working or sleeping. The second thing it does is ventilate the attic, so the air doesn't get so hot. This adds moderated air to the insulation. Fortunately, the generator has enough power to run the fridge, lights, and the fan. If we add a couple of small electric fans, we can get along fairly well, which is why we choose to headquarter at the grove house after storms when the power is out instead of more modern dwellings whose comfort levels rely on air conditioning. Also, the grove house has a propane water heater, stove, and oven. The fan is particularly important during hurricanes because all the windows are boarded up so we can't open them to get cross ventilation in the house which was the primary way of keeping cool.

One thing that happened is that Cindy and Brenda decided to clean things up on Wednesday, and ran the water hose for a couple of hours. Apparently, the water pump didn't like working that hard on the generator power, and the 220 side of the pump starter burned out and it quit working. James figured out how to re-wire it to 110 using the end of an extension cord, and plugged it in to a nearby outlet. The problem is that the 110 starter draws more current than the 220, so when the water pump kicked on everything else browned out and the breaker on the generator input would throw. We found this out while trying to get computers up and running, and every time it happened the computers would crash. The water source at the grove is an artesian well, so there's a little pressure even without the pump, and we finally came up with a system whereby we would throw the breaker to the water pump all day, with the artesian well supplying enough pressure to wash dishes and flush, and at night at shower time we would turn everything else off, start the water pump, take our showers, then turn off the pump and re-start the lights and fan. It wasn't too bad once everyone learned the rules.

Emory came down from Citra with the fruit truck loaded with drums of gasoline and coolers of ice. We use big coolers to transport orange juice from Vero Beach up to Citra, so he went to the Gainesville Ice Company and bought 200-lb blocks of ice and packed them in the coolers. So, we should have enough gas and ice to last for awhile. We kept enough ice to take care of the Arapaho families, then we brought coolers to our friends and customers who were trying to field crews to deal with the damage in the groves. As the ice melts, we put gallon jugs of drinking water and Gatorade right in the coolers—which is a real treat.

Thursday we got the computer server and my desktop machine running, and Jan came in to do payroll. She had gotten in from North Carolina the night before, after having been caught in bad weather when Frances blew through up there. We couldn't do the payroll and billing at the same time, so we just got enough information in to write checks. We had phones at the office the first few days, but we lost them Thursday night. If we don't get them back by Monday, we may have to arrange to call forward the main line to cellphones.

On Thursday, I also worked with the Tropicana people trying to come up with a way to dispose of their wastewater without electricity to run the pumps. Florida Power and Light said that they weren't even going to try and get power out to the groves and rural areas until all the city dwellers had their power restored, so it looks like a wait. We came up with a strategy that will take about a week to become fully operational, but at least we have a plan and everything should be OK. We may also try a couple of interim measures just in case, and Danny and Bobby started working on that Friday and Saturday.

By Friday, we pretty well had everything drained except Glades, where one pump was out of service. We're running the smaller pump at Glades 24/7 with one of our tractors, since there is no power, and making some headway. There's a lot of water, though, and Ivan may dump some more.

We're fortunate in that I design all of our drainage pumps to run with either electricity or tractor power. Where possible, I prefer electric motors for normal operation since the motors don't require much maintenance and it allows automation. For storms, though, the pumps have a power takeoff input splined shaft, so we can back tractors up to them and run without electricity. Since the tractors aren't working when the groves are flooded, they're available. It's a system that has served well over the years, and proved itself again this summer.

There is an immediate need to harvest the lemons. The trees are broken up pretty bad, and won't hold the large crop very long. The fruit is ripe and ready for harvest. At Shadow Lakes, we'll actually have to have a crew with the picking crew to chain saw out broken limbs so that the pickers can work down the tree rows. Unfortunately, all four packinghouses in the area, that can pack lemons, sustained damage and can't run. Plus, they don't have power even if they could run. On top of that, the lemon fruit was pretty well scratched up by the thorns, so they may not be packable for fresh. If we can't resolve these issues, we'll probably try to send them to a processor next week, at a significantly lower price than fresh. Hopefully, crop insurance will cover some of the difference for the Shadow Lakes lemons. There is no crop insurance program for lemons in St. Lucie County, though, so my lemons aren't insured.

If there is a sense of peace amongst all this chaos, it has to be coming from faith. Everything looks bad. However, I have the strong feeling that we will find a way to get through all of it, and there is no logical reason that this could be possible, so it has to be coming from pure faith. The challenge is to stay open, and be able to see the opportunities hidden in the chaos. They have to be there, and I believe that me, Cindy, and the wonderful group of people who work for me will somehow come through this stronger and wiser, and that is the blessing. I keep looking for that—I don't believe that this can all be strictly a negative experience. We have to be careful to make sure we allow the blessings and positive experiences to reveal themselves. They are there, and it's up to us to see them and grow as a result.

Tuesday, September 21, 2004. Big day today. We got our cable TV back. That completes the full complement of civilized amenities. That's essentially 18 days of deprivation. The lack of cable was an inconvenience—nothing more—with the biggest loss being weather reports. I found that the internet just didn't replace TV weather people—the lack of discussion was a handicap in trying to determine what the weather was doing. I could track the hurricanes OK, but it was hard to discern the normal weather outlook without somebody pointing at the map and explaining what was going to happen. We did manage to fill the cable gap by playing nearly our entire collection of videotapes. Had some really neat treasures in there that we hadn't seen for years. So, the time was not entirely wasted. Of course, we had to deal with the weather even if we didn't know what to expect. Over the last two days, we've had 6-12" of rain from feeder bands from Hurricane Ivan that kept swinging over the state, which is now the third hurricane to affect us in the last month. The groves are now actually as wet as or wetter than they were during Hurricane Frances. They were wet to begin with, so all the rain just stacked up. The water furrows are flooded, and rotting fruit is floating all over the place. As we're pumping down the water, the fruit gets caught in the drain pipes at the ends of rows, so the crews have to get down in the water with hoses from gasoline-powered jet pumps to clear the pipes. This is not exactly the best job in the world with all the gooey mess of rotten fruit floating around.

I'm grateful for the rain, though. One of my concerns was that Frances deposited an enormous amount of salt on the trees and in the soil. A lot of the precipitation in the storm was seawater. Many of the people who rode out the storm here mentioned that as the storm came through, the sky at night had a green glow to it. This is a common phenomenon in hurricanes that are coming in from the ocean—the green glow is the phosphorescence from the diatoms in the seawater that fluoresce green when they are disturbed. If the air was full of diatoms, it was also full of seawater. So, the 10-12" of precipitation we received during the storm had a heavy salt load. Fortunately, these heavy rains should have cleansed the trees and flushed much of the salt out of the soil.

The citrus industry in our area is in total shock. More fruit is falling from the trees every day. There are actually three sets of dropped fruit on the ground right now—the fruit that was blown off as the storm went through, the fruit that fell shortly thereafter because the stems were twisted and damaged, and now the fruit that was injured to the point where it cannot heal, so the tree is shedding it off. I would say that the majority of the drop is over, but there will be more since the growers in the path of Charley, which went through a couple of weeks before Frances, are saying that fruit is still dropping over there.

Conservatively, 70% of the Indian River fresh citrus is on the ground. Grapefruit were hit the worst, with the loss probably being closer to 80% in some places. Specialties such as tangerines lost less, but it's still significant. Navels were also hit hard, plus they were scratched and banged up more than other varieties since they are more sensitive to rough handling.

The thing that is occurring, however, is that almost every commercial packinghouse is planning to open and run. There is nowhere near enough fruit to supply all of them, however. That means that a train wreck is coming unless some of them make a different choice. When I bring this up, the universal answer is that they believe that they will find enough fruit. I don't see how it's going to happen, though, so we are being very careful in our choice of packinghouses. Instead of the price, I'm now asking for their fruit purchasing plan. The prices have all gone up, and I presume that those with a good purchasing plan assume they will meet the price. If a packinghouse is making lowball, or outrageously high offers in hopes of getting lucky, I also presume they will probably crash at some point.

Labor is a major issue. We weren't harvesting when the storm hit, so there weren't many migrants here. The local picking crews that stay year-round were immediately snatched up by FEMA-paid debris pickup contractors, who are paying the workers $10/hour in cash. We have lemons that we need to get harvested, but can't find any pickers. Lemons are a minor crop—we're one of only 4 growers in the state that have them—and fresh lemons have to be clipped rather than picked because they are very delicate. Also, we have to wait each day to start harvesting until the turgor pressure of the oil cells in the peel drops to a certain point or the oil cells will be ruptured by mere finger pressure and oil spotting will occur that precludes the use of the fruit for packing. On an average day, a picker can clip about 15-20 boxes of lemons after the time the turgor pressure reaches the harvest point, which is about noon on an average day. We usually pay them about $2.50/box, so they make about $40-50 for about 4 hours work. Since normally there's nothing else for them to do, they go for that deal. Now, however, they want to be guaranteed $90-100 a day, since they can make that much picking up debris. We can't start harvesting any earlier in the day, and they can't pick any faster than they normally do, so that cost would need to be applied to the 15-20 boxes, which would run up the cost to $4-5.00 per box. The cost for loading the fruit is based on the picking price, so that goes up, too. The normal rules are simply out the window. We are working every day to come up with creative solutions to overcome these things, but every time we turn around there's a new challenge.

We're hoping that the federal government will provide some disaster money to supplement our crop insurance for lost fruit. We heard from our Congressional delegation yesterday, though, that the bill to provide the disaster assistance has been sidetracked due to other states wanting to include disaster assistance for their disasters. The states out west want assistance for a drought. Ivan has gone through and caused a lot of flooding all the way up the country—my brother said it flooded out a lot of people in Pennsylvania. The best one is that the State of California is claiming that Ronald Reagan's funeral was a disaster and reimbursement should be included with the bill. It's now 4 times what Florida requested, and growing every day. There's some money available that doesn't require congressional authorization, but it's not enough to bail us out. Fortunately, we have a very strong position—the President's brother is Florida's governor, and one of our Representatives is the Chairman of the Appropriations Committee, which doles out money. They are all doing their best to shake something loose, so there's hope.

We have a couple of employees that are still dealing with dwelling issues. One that qualifies for FEMA assistance ended up getting the Army Corps of Engineers to put a blue tarp on his roof. One had his home destroyed, so he has to find another place to live.

Mold is a big issue. Apparently, mold has an incubation period of 72 hours at 78 degrees. A lot of the houses and buildings with roof damage got wet, and with no power there was no way to control the temperature and humidity, so mold started growing. The only solution in a lot of cases is to rip out all the carpet and drywall, treat the studs and floor with bleach, and rebuild. People can't live there while all that is happening, so many dwellings have been classified as uninhabitable. All this has to happen after the roof has been fixed, of course. The roofers estimate that it will take 2 years to fix all the roof damage. It's been raining for the past two days, so the buildings are getting wetter. In the paper, the School Board estimated that mold has grown on about 40,000 pieces of furniture in the schools, and they will all have to be wiped down with disinfectant before school can resume. They can't start wiping them down until the roofs and air conditioners are fixed. A lot of the air conditioners were on the roofs, and blew off in the storm when the roof let go. Kids have not been in school since the storm, so parents have to deal with day care and such, but they missed work, too, so budgets are stretched thin.

Stress, needless to say, is prevalent. Cindy coined the phrase "melt down" to describe the point where people finally get overwhelmed and just start freaking out instead of maintaining a rational approach to problem solving. Melt down has a myriad of manifestations. People are cross and grouchy, attention spans are limited, and the smallest problem seems insurmountable. We're spending a lot of time trying to keep people focused on the most important issues, and crossing problems off the list one at a time. At this point, it's hard to see how we'll ever get ahead, so it's back to the faith thing. We just have to operate based on the faith that if we keep putting one foot in front of the other, and making the best choices we can, that at some point we'll start to make discernible forward progress. Part of the melt down might be impatience. We want to get things settled as soon as possible, but we keep getting confronted with obstacles such as heavy rains, lack of pickers, closed schools, etc., etc., etc. It takes a conscious effort to be patient, and trust that we will at some point be able to put this all behind us.

HCCGA Grower Newsletter, September 8, 2004

As Hurricane Frances moved west and north away from the central Florida groves, Dennis Broadaway, Executive Vice President of the Haines City Citrus Growers Association, expressed the views of many fresh fruit cooperatives.

Wow, what a month. Two hurricanes crossing the state within four weeks of each other, with Polk County at the crossroads. And another hurricane on the horizon at the time of this writing. Between Charley, Frances, and now potentially Ivan, we can look to Dr. Attaway to write a sequel to his book *Hurricanes and Florida Agriculture*, which he published in 1999.

Damage to trees and fruit from both Charley and Frances was both widespread and variable, depending on the location of groves in relation to each storm's path. About the only citrus growing area in the state that was spared the fury of each of these hurricanes appears to be the Lee, Collier, Hendry county area of southwest Florida. I base that observation on the post-hurricane condition of the CPI groves located in these counties as reported by Mr. Mike Murphy, the newly appointed general manager of those project groves. He reports very little to insignificant fruit loss or damage in the three project groves.

With regard to HCCGA member groves further north, Hurricane Charley put varying amounts of fruit on the ground and broke varying amounts of trees in those that were in the direct path, which started out at about 30 miles wide in DeSoto County and into Hardee County, and ended up at about 15 miles wide in Polk County. In Hardee County, our member groves, which are made up of mostly round oranges, lost up to 50% of their fruit, and sustained damage to old, weaker trees. In Polk County, those groves in the path of the storm lost up to 70% of their grapefruit, up to 30% of their oranges, and up to 15% of their tangerines. Now that's not to say every grove of those varieties sustained that level of fruit loss. Those are maximums. There are many, many groves that suffered less fruit loss. In all cases, most of the fallen fruit came from the tops of the trees.

Sunburst tangerine trees, which are known to be "brittle", sustained the largest amount of tree damage as a result of limbs loaded with a large crop breaking near the trunk. Again, the level of damage varied significantly from grove to grove in the path of Charley.

Although we did not have the 100 to 115 mph winds in Frances that we experienced in Charley, there was additional fruit drop as a result of the many hours of relentless 40 to 60 mph winds experienced in Frances. We are still in the first days of evaluating fruit loss from Frances, but at this time fruit loss appears to be less than expected. But it's still early. In our observations of fruit drop resulting from Charley over these past three weeks, we saw fruit continue to drop well into the second and third weeks. So time will tell the extent of fruit loss.

Both hurricanes scarred a lot of fruit in their path. Everything from limb rubs and punctures to fruit throughout the tree to sandblasting of fruit on lower limbs will most certainly adversely impact fresh fruit external quality for the coming season. Again, the level of damage is specific to fruit variety and location, with some varieties showing little external fruit damage, and others showing higher levels at this time.

We have had crop and tree insurance adjusters in those member groves that had purchased insurance recording their preliminary observations. Because our insured member crops are insured as fresh fruit, the final tabulation of loss by the insurance adjusters will not take place until the fruit is harvested to determine the fresh fruit marketability of the crop.

Florida Citrus Mutual, working on behalf of the growers, has received a commitment of $225 million in hurricane relief from the federal government that will be available to growers based on the level of hurricane crop damage. The details have not been worked out as of yet, but it appears growers will be receiving some level of federal relief dollars in the future.

Facilities and equipment at both HCCGA locations came through both hurricanes with relatively minor damage. The small structure enclosing the boiler will have to be replaced, along with a roof over a porch area. Other than that, we sustained minor damage to the packinghouse and main office. A large roll-away door blew in at the production/harvesting shop facility, allowing the wind to blow off a section of the roof. In addition, the top blew off of the water tank used for fire protection at that facility. We are still in the process of getting appraisals from contractors for repairs.

So, with the 2004-05 season about to begin, what can we expect for the coming year? Typically limited supplies of fruit equates to higher prices. This should most certainly be realized in the fresh fruit markets, as supplies will be impacted by not only the loss of fruit that drops to the ground, but also by the lesser quantity of fruit still on the tree that meets external fresh fruit quality standards. Processed fruit should also realize an increase in prices, but maybe not to the extent of fresh fruit due to the current large world-wide inventory of orange juice. We can only hope these increases in prices are enough to offset the grower's loss of yield of marketable fresh fruit and of juice fruit.

USDA Research Laboratory in Ft. Pierce, Florida

The USDA research laboratory in Ft. Pierce was hit hard by the 2004 hurricanes. Laboratory Director Dr. Calvin E. Arnold writes as follows:

The U.S. Horticultural Research Laboratory was in the direct path of both hurricanes Frances and Jeanne with only a brief three week recovery time between them. The major damage was to our 36 research greenhouses which house a significant portion of our valuable citrus germplasm used to develop new varieties for the U.S. citrus industry. Fortunately, the citrus tree is not as fragile as many other horticultural crops, i.e. floral crops and vegetables. Thus, we lost only a limited amount of germplasm even though the potted citrus trees were piled in the corners of the greenhouses mingled with rubble.

Repair of the damaged greenhouses was a major undertaking costing more than $1 million. However, our lab was fortunate that the recovery funds were provided by our USDA headquarters in D.C. Repair of the greenhouses was completed in approximately May 15, 2005. Our research grove in Ft. Pierce sustained minimal damage to the buildings, but about 20% of the trees had scaffold limbs broken and/or blown over. However, less than 5% of the trees were actually lost. After a lot of labor by our farm crew, the trees have been trimmed and restored to their upright position. Our breeders were able to make some valuable observations about differences among the selections in respect to their ability to withstand hurricane winds.

Indian River Research and Education Center

Dr. Brian Scully, Director of the Indian River Research and Education Center, said that everyone of the researchers at the Center in Ft. Pierce were set back by damage from Hurricane Frances, and the Center buildings received varying amounts of damage, particularly roof damage. Leaky roofs were a problem for the researchers and support staff.

Frances also did major damage to the 120 acre orange and grapefruit grove at the Center which was 5-6 years old and just beginning to bear.

Reports from Indian River Area Growers and Packers

Jeff Cusson is President and R. Scott Hurley is Director of Fresh Fruit for Becker Holding Corporation on South Jenkins Road in Fort Pierce, Florida. In a discussion with them at the Indian River Citrus Seminar they indicated that before the ravages of citrus canker, Becker had 15,000 acres of citrus and was one of the largest citrus growers in the state. As a result of canker their acreage is somewhat lower. Becker's northern most grove is off Highway 60 about 4 miles west of I-95. Their southern most grove is off Ridge Road (State Road 708) near Hobe Sound. The distance from north to south is 50 miles. A third of the company acreage is grapefruit and two-thirds are early and Valencia oranges which are processed.

Although there was an estimated 30% loss, Scott indicated that the company "fared OK" from Frances as there had been an oversupply of fruit, particularly grapefruit.

The groves were heavily battered by Frances as the hurricane sat off shore for a full day before moving slowly across the Indian River area. There was significant damage to infrastructure, particularly barns, pump houses and greenhouses which were totaled. Roof damage was rampant.

Gregory P. Nelson is President of DNE World Fruit Sales in Fort Pierce, Florida. DNE is a major grower and packer of grapefruit, oranges and specialty fruit both from its own groves as well as for other companies. DNE groves are principally located in Indian River, St. Lucie and Martin counties on the east coast and Hendry County in southwest Florida, and felt the wrath of both Hurricanes Frances and Jeanne in 2004 and Hurricane Wilma in 2005.

Mr. Nelson indicates that tree loss was not a major problem from Hurricane Frances, probably less than 3%. They had significant fruit loss but as Hurricane Jeanne followed so closely behind Hurricane Frances and both followed the same course, it would be difficult to separate the losses from the two hurricanes. Punctured fruit was a problem in that it would hold to the trees for a time, but later drop on the ground and was not usable.

While the National Hurricane Center classified Hurricane Frances as a category 2, the damage was greater than would have been anticipated as wind gusts were in excess of 100 mph and small tornadoes were frequent.

Mr. Nelson was pleased with how well the company's Fallglo tangerines held to the trees. Droppage after the hurricane was not nearly as severe as other varieties.

Rainfall from Hurricane Frances was very heavy, particularly in Martin and St. Lucie counties, with many locations receiving 8-10 inches. This was further complicated by the remnants of Hurricane Ivan which drifted south from Virginia and across Florida from east to west depositing another eight inches or more of rainfall in St. Lucie, Martin, Okeechobee and adjoining counties.

TABLE 2-2. WATCH AND WARNING SUMMARY FOR HURRICANE FRANCES DURING THE PERIOD SEPTEMBER 1-7, 2004 AS IT MOVED FROM THE CENTRAL BAHAMAS AND CROSSED THE FLORIDA PENINSULA.

Date	Time (GMT)	Action	Location
9/1	1500	Hurricane warning issued	Central Bahamas
9/1	1500	Hurricane watch issued	Northwestern Bahamas
9/1	0000	Hurricane warning issued	Northwestern Bahamas
9/2	0300	Hurricane watch issued	FL east coast from Florida City to Flagler Beach including Lake Okeechobee
9/2	0300	Tropical storm watch issued	Florida Keys from Florida City to Seven Mile Bridge including Florida Bay
9/2	0900	Hurricane watch extended southward	Florida Keys from Florida City to Craig Key
9/2	1500	Hurricane warning issued	FL east coast from Florida City to Flagler Beach including Lake Okeechobee
9/2	1500	Hurricane watch and tropical storm warning issued	Florida Keys from Florida City to Seven Mile Bridge including Florida Bay
9/3	1500	Hurricane watch extended northward	FL coast from Flagler Beach to Fernandina Beach
9/3	2100	Tropical storm warning issued	FL coast from Flagler Beach To Fernandina Beach
9/4	0300	Tropical storm warning issued	FL southwest coast from Bonita Beach to Florida City
9/4	0900	Tropical storm warning extended northward	FL west coast from Bonita Beach to Englewood
9/4	1500	Tropical storm warning extended northward	FL and GA coasts from Fernandina Beach to Altamaha Sound
9/4	1500	Tropical storm warning extended northward	FL west coast from Englewood to Anna Maria Island
9/4	1500	Tropical storm watch extended northward	FL west coast from Suwannee River to St. Marks
9/4	2100	Tropical storm warning extended northward	FL west coast from Anna Maria Island to St. Marks
9/4	2100	Tropical storm watch extended westward	FL Gulf coast from St. Marks to Panama City
9/4	2100	Hurricane warning discontinued	Northwestern Bahamas except Grand Bahama, Abaco, Bimini, and Berry Island
9/5	0900	Hurricane warning changed to tropical storm warning	FL Southeast coast from Deerfield Beach to Florida City
9/5	0900	Tropical storm warning issued	Florida Keys south of Seven Mile Bridge including the Dry Tortugas
9/5	1500	Hurricane warning issued	FL Gulf coast from Suwannee River to Destin
9/5	1500	Hurricane watch discontinued	FL east coast from Flagler Beach to Fernandina Beach
9/5	1500	All warning discontinued	Northwestern Bahamas
9/5	2100	Hurricane warning extended southward	FL west coast from Suwannee River to Anna Maria Island
9/5	2100	Hurricane warning changed to tropical storm warning	FL east coast from Jupiter Inlet to Flagler Beach including Lake Okeechobee
9/5	2100	All warnings discontinued	FL east coast from Florida City to Jupiter Inlet
9/6	0600	Tropical storm warning discontinued	FL west coast from Bonita Beach to Florida City including the Florida Keys, the Dry Tortugas, and Lake Okeechobee
9/6	0600	All warnings discontinued	FL west coast south of Englewood and FL east coast south of Cocoa Beach
9/6	1500	Hurricane warning changed to tropical storm warning	FL Gulf coast form Indian Pass to Destin and FL west coast from Anna Maria Island to Suwannee River
9/6	1500	All warnings discontinued	GA coast, FL east coast, and FL west coast south of Anna Maria Island
9/6	2100	Hurricane warning changed to tropical storm warning	FL Gulf coast from Suwannee River to St. Marks
9/6	2100	All warnings discontinued	FL coast west of St. Marks and south of Suwannee River
9/7	0300	All warnings discontinued	Florida coast

Source: John L. Beven II, National Hurricane Center. 17 December 2004.

George Hamner, Jr. is President of Indian River Exchange Packers, Inc. in Vero Beach Florida. Although the company offices are in Indian River County, their 1700 acres of groves are in St. Lucie county. During Frances the family and staff rode out the hurricane in the heavy duty office building on SW 9th Street in Vero Beach. It is the company shelter when major hurricanes strike the area.

Frances was a slow moving hurricane that required a day to pass. George described the effect on the packinghouse as "we lost one-fourth of the roof which peeled back like a sardine can." During the storm we had put 22 cars and a boat into the packinghouse, but luckily none were in the path of the collapsing roof. In the groves, however, the wind gusts devastated the barns and sheds and blew the tops off the pumping stations.

The first variety which Indian River Exchange Packers (IREP) pack each year is lemons from the Big Cypress Reservation owned by the Seminole Indian Tribe. IREP is one of two Florida houses to handle lemons. Starting in late July and early August, the company had a house full of lemons. "Frances hit on Saturday and our insurance man was here on Monday. We ordered the roof and by working 20 hour days we had it in place and ready to go the following week. Unfortunately, Hurricane Jeanne hit on Saturday and we never packed a lemon or anything else in our house in the 2004/05 season."

Inland Counties

John Jackson, County Extension Agent IV—Citrus, reports that as inland counties did not receive hurricane force winds from Frances, damage in Lake and Orange counties was not severe. However, as Brevard and Osceola counties were nearer to the hurricane center, some fruit was on the ground, particularly grapefruit, but tree damage was restricted to weak trees with foot rot or tristeza. This was also true of groves in eastern Seminole County. All counties received heavy rainfall.

CHAPTER 3

HURRICANE IVAN, SEPTEMBER 5-24, 2004

Ivan was a classical, long-lived Cape Verde hurricane that reached Category 5 strength three times on the Saffir-Simpson Hurricane Scale (SSHS). It was also the strongest hurricane on record that far south east of the Lesser Antilles. Ivan caused considerable damage and loss of life as it passed through the Caribbean Sea.—Stacy R. Stewart, National Hurricane Center, 16 December 2004.

Ivan was the only one of the four major Florida hurricanes in 2004 which did not make a direct hit on the Florida citrus industry. However, as the remnants of Ivan drifted back to the south it crossed Florida from the Atlantic to the Gulf accompanied by heavy rainfall, particularly in St. Lucie and Martin counties.

On September 2, 2004, Florida growers were too worried about giant Hurricane Frances and her threat to the Florida east coast to notice the development of a new tropical depression at 9.70°N latitude, 27.6°W longitude in the eastern Atlantic ocean. Despite the low latitude, the depression developed rapidly into a tropical storm which was assigned the name Ivan.

As Ivan moved to the West, still below 10°N latitude, it attained hurricane strength at 0600 UTC on September 5, 2004, 1,000 miles east of Tobago in the Windward Islands, but Florida growers hardly noticed as Frances was bearing down on the grapefruit groves of Martin, St. Lucie and Indian River counties. One who noticed was Chet Townsend, publisher of *UltimateCitrus.com Citrus News* who wrote that "Another concern for Florida growers is Hurricane Ivan which is a week away from affecting Florida again."

On September 6, as Frances slowly lumbered across the Florida Peninsula and electricity went off from Jacksonville to West Palm Beach, Ivan intensified to category 4 on the Saffir-Simpson scale and became the southern most major hurricane on record.

Now a major threat to the islands of the Caribbean, Ivan was closely monitored by U.S. Air Force Reserve reconnaissance aircraft. Approaching the southern Windward Islands, the hurricane experienced periods of weakening and strengthening before it passed six miles southeast of Granada as a category 3 hurricane. Then

ABLE 3-1. WATCH AND WARNING SUMMARY FOR HURRICANE IVAN, SEPTEMBER 14-24, 2004, AS IT STRUCK THE FLORIDA PANHANDLE AND LATER LOOPED BACK OVER SOUTHEAST FLORIDA.

Date	Time	Action	Location
9/14	0000	Hurricane warning adjusted	Havana east to Pinar Del Rio and Isle of Youth
9/14	0300	Hurricane watch issued	Morgan City, LA to St. Marks, FL
9/14	1500	Tropical storm warning issued	Tulum to Progreso (Yucatan)
9/14	1800	Tropical storm warning issued	Havana east to Pinar Del Rio and Isle of Youth
9/14	2100	Hurricane warning issued	Grand Isle, LA to Apalachicola, FL
9/14	2100	Hurricane watch issued	Morgan City, LA to Grand Isle, LA
9/14	2100	Tropical storm warning issued	Intracostal City, LA to Grand Isle, LA, Apalachicola, FL to Yankeetown, FL
9/14	2100	All watches/warnings discontinued	Yucatan, Cuba
9/15	0300	Tropical storm watch discontinued	Seven Mile Bridge to Dry Tortugas
9/15	2100	Tropical storm warning discontinued	Morgan City, LA to Intracostal City, LA
9/16	0900	All watches and warnings discontinued	West of Grand Isle, LA
9/16	0900	Tropical storm warning issued	Mouth of Pearl River, LA to Grand Isle, LA
9/16	1200	Tropical storm warning discontinued	Apalachicola, FL to Yankeetown, FL
9/16	1500	Tropical storm warning issued	Mouth of Pearl River, LA to Apalachicola, FL
9/16	1500	Tropical storm warning discontinued	Grand Isle, LA to mouth of Pearl River, LA
9/16	1800	Tropical storm warning discontinued	West of MS/LA Border
9/16	2100	Tropical storm warning discontinued	MS/LA Border to Apalachicola, FL
9/22	2300	Tropical storm warning issued	Mouth of MS River to Sargent, TX
9/23	1500	Tropical storm warning discontinued	Mouth of MS River to Morgan City, LA
9/23	2100	Tropical storm warning discontinued	San Luis Pass, TX to Sargent, TX
9/24	0300	All warnings discontinued	Remaining Coastal (San Luis Pass, TX to Morgan City, LA)

Source: Stacy R. Stewart, National Hurricane Center, Miami, FL. 16 Dec 2004. (Updated 27 May 2005.)

on September 8 Ivan moved into the Caribbean Sea with winds of 160 mph, category 5, after which it briefly weakened slightly to category 4 as it passed south of Jamaica and made a westward turn re-intensifying to category 5. Chet Townsend noted that "Florida was a likely target," but as the hurricane passed south of Grand Cayman Island on September 12 and maintained a westerly course towards the Yucatan Channel, the threat to the citrus industry lessened.

Helped by the warm water in the western Caribbean Sea and the Yucatan Channel, Ivan maintained its category 5 status for 30 hours as it entered the southern Gulf of Mexico early on September 14 and turned to the north-northwest and then north toward Alabama and the Florida Panhandle. The Florida citrus grower was finally able to heave a sigh of relief as Mobile and Pensacola prepared for the worst.

Ivan made landfall just west of Gulf Shores, Alabama on September 16 as a category 3 hurricane with 120 mph winds. Moving across the barrier islands of Alabama, the hurricane turned to the north-northeast and weakened into a tropical storm over central Alabama. Gradually turning to the northeast, Ivan became a tropical depression over northeast Alabama. Continuing northeast for 36 hours Ivan merged with a frontal system and became an extra-tropical low over the Del Marva peninsula on September 18. The National Hurricane Center noted that "even as an extra-tropical low, the remnant circulation of Ivan was identifiable in both surface and upper-air data." Over the next three days, the low moved south and southwest and crossed the southern Florida peninsula on September 21 depositing heavy rain accompanied by flash floods. Precipitation in Martin, St. Lucie and Palm Beach counties exceeded seven inches.

After emerging into the warm waters of the Gulf of Mexico, Ivan moved west and developed a low-level circulation center and became a tropical depression on September 22. It later regained tropical storm strength, turned to the northwest and finally made a second landfall over southwestern Louisiana as a tropical depression.

In All, Ivan existed for 22.5 days over a course of 5,600 miles.

CHAPTER 4

HURRICANE JEANNE, SEPTEMBER 13-28, 2004

Jeanne produced heavy rain over Guadeloupe, Puerto Rico and the Dominican Republic and caused an estimated 2000 or more deaths in Haiti from torrential rainfall flooding. Finally, Jeanne hit the northern Bahamas and then the central Florida east coast as a category three hurricane.—M. B. Lawrence and Hugh D. Cobb, National Hurricane Center, January 7, 2005.

The Meteorology

The system which developed into Hurricane Jeanne began as a tropical wave which moved off the west coast of Africa into the eastern Atlantic Ocean on September 7, 2004. This was two weeks after a similar system which gave birth to Hurricane Frances, emerged off the African Coast on August 21, 2004, and 5 weeks after the tropical wave which became Hurricane Charley had emerged from the African Coast on August 4, 2004. Three Cape Verde hurricanes which were destined to damage the Florida citrus industry had formed over a six week period.

The tropical wave moved slowly west-northwest over the eastern Atlantic becoming tropical depression eleven 70 miles east-southeast of Guadeloupe on the evening of September 13 and reaching tropical storm status as it reached the Leeward Islands on the 14th of September. Gradually strengthening, Jeanne moved south of the Virgin Islands and made landfall in Puerto Rico on the 15th. She then moved into the Mona Passage where her winds reached minimal hurricane force. After making landfall over the Dominican Republic on September 17, Jeanne weakened to a depression and moved into the Atlantic Ocean where her storm center dissipated on the 18th. Jeanne's slow motion as she crossed Hispaniola had produced heavy rains, floods and mud slides with thousands killed in Haiti.

However, Jeanne was not finished. A new center formed well to the northeast of the old center, and influenced by Hurricane Ivan, she floundered in a weak steering flow for 5 days as she moved north over the southeastern Bahamas as a tropical storm. Making a complete loop about 500 nautical miles east of the northwestern Bahamas, Jeanne gradually strengthened to hurricane force on September 20 with 90+ mile per hour winds. Completing her loop on September 23, she began a westward move over Abaco and Grand Bahama Island on September 25 with winds increasing to 120 mph, Category 3 on the Saffir-Simpson Scale.

The 50 nautical mile wide eye of Jeanne, closely following the same path as Frances, finally made landfall on September 26, at the southern end of Hutchinson Island just east of Stuart. This was only two miles from Sewall's Point where the center of Hurricane Frances had come ashore three weeks earlier. Maximum winds at landfall were estimated to be 120 mph over a small area just north of the center.

Jeanne created heavy damage in the citrus groves of the Indian River area from Palm Beach to Brevard County before weakening as she moved west and became the third hurricane to pass through Polk County in a six week period. Polk County became the only Florida county to feel the eyewall of all three hurricanes, Charley, Frances and Jeanne. After recurving 30 miles north of Tampa, Jeanne diminished to tropical storm status. Damage in central Florida groves was significant, particularly in Okeechobee, Highlands, Hardee and Polk counties, but was much less than in the vulnerable east coast counties.

During Jeanne's passage over Florida, a belt of hurricane force winds 90 miles wide had affected the Florida East Coast from the Cape Canaveral area south to near Stuart. The highest reported wind gusts were 128 mph at Fort Pierce Inlet and 122 mph at Vero Beach. Sustained winds of hurricane force spread inland half way across the state and tropical storm force winds affected much of the remainder of central Florida.

Rainfall up to 8 inches accompanied Jeanne as she crossed the Florida Peninsula, with a heavier 11-13 inch rain band observed over Osceola, Broward and Indian River Counties.

Wind speeds and rainfall at selected Florida locations follow in Table 4-1. The Watch and Warning Summary for key actions by the National Hurricane Center during the approach of Hurricane Jeanne to the east coast of Florida and the subsequent passage of the hurricane across Florida are shown in Table 4-2.

DNE World Fruit Sales

Gregory P. Nelson is President of DNE World Fruit Sales in Fort Pierce, Florida. DNE is a major grower and packer of grapefruit, oranges and specialty fruit from its own groves as well as others. DNE groves are principally located in Indian River, St. Lucie and Martin counties on the east coast and Hendry County in southwest Florida.

TABLE 4-1. WINDSPEEDS AND RAINFALL FOR HURRICANE JEANNE.

Location	Sustained winds (mph)	Peak Gusts (mph)	Rainfall (in.)
Daytona Beach	52	63	2.31
Fort Lauderdale	40	56	0.74
Fort Pierce	52	88	—
Gainesville	44	60	4.90
Melbourne	60	78	6.04
Orlando	61	77	5.40
Palm Beach	—	69	9.10
Sanford	53	69	4.37
Unofficial Observations			
Fort Pierce Inlet	—	128	—
Jensen Beach	—	105	—
Kenansville (FAWN)	—	—	11.97
Lakeland (South)	—	82	8.09
Live Oak	—	—	10.88
NASA Wind Tower (23.53°N, 80.77°W)	58	83	—
NWS Melbourne	91	—	6.12
Port Canaveral USCG	87	—	—
Port St. Lucie	105	—	—
Sebastian Coastal Monitoring System	82	116	—
Vero Beach	78	122	—
Wimauma	55	81	2.58

Source: M. B. Lawrence and Hugh D. Cobb. National Hurricane Center, Miami, FL. January 7, 2005.

DNE suffered damage from Hurricane Jeanne both to groves and to the St. Lucie packinghouse. Damage by Jeanne was more severe than Hurricane Frances due to higher sustained winds and more frequent tornadoes. The damage was most severe in Martin County and the southern parts of St. Lucie County, where heavy fruit losses were sustained and defoliation was heavy. In the hardest hit groves, the bark was occasionally stripped from the trees, something not seen since the severe hurricane which made landfall near Ft. Pierce in 1949. However, the loss of trees was not more than 5%.

Precipitation was a continuing scourge, adding to the previous heavy rains from Hurricane Frances and the backwash from Hurricane Ivan. The total rainfall in St. Lucie county from the three systems was estimated at 25 inches, making it impossible to carry out regular grove caretaking, harvesting and other operations.

"In summarizing grapefruit losses, it was estimated that in 10% of the groves in Martin and the southern part of St. Lucie counties the loss of fruit was so severe that it was not economical to harvest. In the northern part of St. Lucie County and in Indian River County, the loss was less and a greater number of groves were picked."

Indian River and St. Lucie Counties

George Hamner, Jr. is President of Indian River Exchange Packers, Inc., in Vero Beach, Florida. Although the company offices are in Indian River County their 1700 acres are in St. Lucie County. George recalls that after Hurricane Frances the company had a house full of lemons which could not be packed due to damage to a part of the packinghouse roof. Unfortunately, Hurricane Jeanne completed the destruction of the roof which was lifted up by 120 mph gusts causing two walls to collapse. Fortunately there were not any cars under the roof and only one piece of packing equipment was affected, although there were also problems with leaking which caused water damage to the offices.

"We still had a house full of lemons which we could not ship, but other packinghouses such as DNE packed and shipped them for us."

"It was quite a spectacle. Even ABC News set up in our parking lot with the devastated packinghouse as a background. They said that they were tired of using mobile home parks as a background. Figures 4: 1-4 show the heavy damage to the Indian River Exchange Packinghouse."

Becker Holding Corporation

R. Scott Hurley is Director of Fresh Fruit and Jeff Cusson is President of Becker Holding Corporation. Jeff noted that Jeanne was a much stronger hurricane than Frances and knocked off a great deal more fruit. Due to the loss of leaves from Frances there had been no canopy to protect the fruit and as a result they lost 50% of the remaining crop.

TABLE 4-2. WATCH AND WARNING SUMMARY FOR HURRICANE JEANNE DURING THE LOOP OVER THE BAHAMAS AND SUBSEQUENT PATH ACROSS PENINSULA FLORIDA, SEPTEMBER 16-27, 2004.

Date	Time (GMT)	Action	Location
9/16	1500	Hurricane Watch changed to Hurricane Warning	Southeastern Bahamas including Acklins and Crooked Islands, Inaguas, Mayaguana, Ragged Island, Turks and Caicos.
9/16	1500	Hurricane Watch issued	Central Bahamas including Cat Island, Exumas, Long Island, Rum Cay and San Salvador
9/17	2100	Hurricane Warning changed to Tropical Storm Warning	Southeastern Bahamas including Acklins and Crooked Islands, Inaguas, Mayaguana, Ragged Island, Turks and Caicos
9/17	2100	Hurricane Watch changed to Tropical Storm Watch	Central Bahamas including Cat Island, Exumas, Long Island, Rum Cay and San Salvador
9/19	1000	Tropical Storm Warning discontinued	Southeastern Bahamas including Acklins and Crooked Island, Inaguas, Mayaguana, Ragged Island, Turks and Caicos
9/19	1000	Tropical Storm Watch discontinued	Central Bahamas including Cat Island, Exumas, Long Island, Rum Cay and San Salvador
9/23	0900	Tropical Storm Watch issued	Central Bahamas including Cat Island, Exumas, Long Island, Rum Cay and San Salvador
9/23	1730	Hurricane Watch issued	Northwest Bahamas
9/24	0900	Hurricane Warning issued	Northwest Bahamas: Abacos, Andros Island, Berry Islands, Bimini, Eleuthera, Grand Bahama Island and New Providence
9/24	0900	Hurricane Watch issued	Florida City to St. Augustine, FL
9/24	0900	Tropical Storm Warning issued	Central Bahamas: Cat Island, Exumas, Long Island, Rum Cay and San Salvador
9/24	2100	Hurricane Watch changed to Hurricane Warning	Florida City to St. Augustine, Florida including Lake Okeechobee
9/24	2100	Hurricane Watch issued	St. Augustine, Florida to Altamaha Sound, Georgia
9/24	2100	Tropical Storm Watch issued	Florida City to Anclote Key, Florida including the Florida Keys north of the Seven Mile Bridge
9/25	0900	Tropical Storm Watch changed to Tropical Storm Warning	East Cape Sable to Anclote Key, Florida
9/25	0900	Tropical Storm Watch extended northward	Anclote Key to Ochlockonee River, Florida
9/25	1500	Tropical Storm Warning issued	St. Augustine, Florida to Altamaha Sound, Georgia
9/25	1500	Tropical Storm Warning extended northward	Anclote Key to Suwanee River, Florida
9/25	1500	Hurricane Watch issued	Englewood to Suwanee River, Florida
9/25	2100	Tropical Storm Warning extended northward	Suwanee River to Ochlockonee River, Florida
9/25	2100	Tropical Storm Warning discontinued	Central Bahamas: Cat Island, Exumas, Long Island, Rum Cay and San Salvador
9/26	0300	Tropical Storm Warning extended westward	Ochlockonee River to Indian Pass, Florida
9/26	0300	Hurricane Warning changed to Tropical Storm warning	Abacos Island, Berry Islands, Bimini and Grand Bahama Island
9/26	0900	Hurricane Warning issued	Englewood to Suwanee River, Florida
9/26	0900	Tropical Storm Warning extended northward	Indian Pass to Destin, Florida
9/26	0900	Hurricane Warning discontinued	Hallandale to Florida City, Florida
9/26	0900	Tropical Storm Warning discontinued	Florida City to Chokoloskee, Florida including the Florida Keys north of the Seven Mile Bridge
9/26	0900	Tropical Storm Warning discontinued	Northwest Bahamas: Abacos Island, Berry Islands, Bimini and Grand Bahama Island
9/26	1300	Hurricane Warning discontinued	Hallandale to Deerfield Beach, FL
9/26	1500	Hurricane Warning discontinued	Deerfield Beach to Jupiter Inlet, Florida
9/26	1500	Hurricane Watch discontinued	St. Augustine, Florida to Altamaha Sound, Georgia
9/26	1800	Hurricane Warning discontinued	Jupiter Inlet to Cocoa Beach, Florida
9/26	1800	Hurricane Warning changed to Tropical Storm Warning	Cocoa Beach to St. Augustine, Florida including Lake Okeechobee
9/26	1800	Hurricane Warning changed to Tropical Storm Warning	Englewood to Suwanee River, Florida

Source: M. B. Lawrence and Hugh D. Cobb. National Hurricane Center. Miami, FL. 7 Jan 2005.

TABLE 4-2. (CONTINUED) WATCH AND WARNING SUMMARY FOR HURRICANE JEANNE DURING THE LOOP OVER THE BAHAMAS AND SUBSEQUENT PATH ACROSS PENINSULA FLORIDA, SEPTEMBER 16-27, 2004.

Date	Time (GMT)	Action	Location
9/26	1800	Tropical Storm Warning discontinued	Chokoloskee to Bonita Beach, Florida
9/26	1800	Tropical Storm Warning discontinued	Bonita Beach to Englewood, Florida
9/27	0300	Tropical Storm Warning discontinued	Apalachicola to Destin, Florida and Englewood to Anclote Key, Florida
9/27	0900	Tropical Storm Warning extended northward	Altamaha Sound, Georgia to Santee River, South Carolina
9/27	0900	Tropical Storm Warning discontinued	Cocoa Beach to Flagler Beach, Florida
9/27	1500	Tropical Storm Warning discontinued	Anclote Key to Apalachicola, FL
9/27	1800	All coastal warnings discontinued	

Source: M. B. Lawrence and Hugh D. Cobb. National Hurricane Center. Miami, FL. 7 Jan 2005.

It was 4-6 weeks after Jeanne passed before harvesting began. This allowed time to assess the remaining crop and establish a price structure. There was only about a half crop for this year of which lost leaves and tree losses contributed a further 80% loss. The situation was worsened by an estimated 25 inches of rain which resulted from Frances, the backlash from Ivan and Hurricane Jeanne.

Lake County

Charles M. "Charlie" Hendrix, Jr. is a retired veteran of over 50 years service to the Florida citrus processing industry. Charlie was associated for many years with Citrus Central and several other companies. Charlie writes that "Jeanne surprised us in Lake County when she turned toward the north and exposed south Lake County to the northeast quadrant with 70-80 mph winds and gusts to 90 mph for over half an hour. Properties on the north and west side of our big lakes were pounded. We are not big growers in Lake County now like we were in the 1970s when we were second only to Polk County in orange production, but Jeanne wiped out some of the old trees which had survived here and there. When the winds got up, many old oak trees along the lake shores were destroyed, including my own yard where I lost five large trees.

"Jeanne also did damage in Sumter County and Citrus County but not to the extent as in our area as she lost strength over land."

Inland Counties

John Jackson, County Extension Agent IV-Citrus indicates that Hurricane Jeanne was more severe in his counties than Hurricane Frances, but that damage was less than for Hurricane Charley except in Brevard County which is a coastal county. Volusia County is also a coastal county but as Hurricane Charley exited the state near New Smyrna Beach, the damage was greater than from Jeanne. However, as Jeanne took a northwesterly turn toward Sumter and Citrus counties, the western part of Lake County was on the strong easterly side of the hurricane and did receive some winds near hurricane force which blew down many oak trees, but there are not many citrus trees in that area since the 1985 freeze (Attaway, 1997).

Indian River Research and Education Center
Dr. Brian Scully, Director

Frances had done major damage to the greenhouses at the Center. Jeanne destroyed anything which had survived.

Jeanne also did additional damage to the 120 acre orange and grapefruit grove, but we were able to prune and stake the trees and start the grove on the way to recovery. One new greenhouse was built after Jeanne and leaky roofs were repaired.

Kevin Bouffard, writer for The Ledger in Lakeland, Florida does an outstanding job covering Florida agriculture, particularly the Florida citrus industry, for Ledger readers. His daily By-Line is a dependable source of information for Florida growers, packers and processors. The following is a typical column written by Bouffard during the crisis days of the 2004 hurricanes.

Jeanne Squeezes Life Out of Citrus

Haines City—Like a baseball player struggling at the plate, the third strike was the worst. Hurricane Jeanne caused more damage to citrus over a wider area in Polk County than hurricanes Charley and Frances, said Charles Counter, the director of field operations at the Haines City Citrus Growers Association.

"This last storm was pretty democratic—it got everybody pretty equally," said Counter, who manages caretaking for about 4,800 grove acres in Polk. "Some of these groves have lost so much I don't know if there's enough fruit left to pick."

Figures 4: 1-4. Show damage to facilities at Indian River Exchange Packers, a prominent fresh fruit house in Vero Beach, FL. Source: George Hamner, Jr. Indian River Exchange Packers, Vero Beach, FL

A30963

Figure 4-5. Shows Indian River Exchange Packers as rebuilt after Hurricane Jeanne. Source: George Hamner, Jr. Indian River Exchange Packers, Vero Beach, FL.

Charley and Frances destroyed an estimated 40 percent of the citrus crop in Polk, Florida's largest citrus-producing county.

But Charley affected only a narrow strip along the path from Fort Meade through Lake Wales to Haines City, Counter said. Frances was a bigger storm, but its winds had died down considerably by the time it reached Polk.

Jeanne blew across most of the county with sustained winds between 90 mph and 100 mph. Counter estimated Jeanne left his groves with as little as 25 percent of the grapefruit crop remaining on the trees and about half the Navel oranges, primarily a fresh crop.

Other oranges, used mostly in juice, fared better with about 65 percent of that crop remaining, Counter said. About 70 percent of Polk's tangerine crop remained on the trees in Jeanne's wake.

Garvie Hall, who manages more than 200 citrus acres in the Bartow area, estimated he lost about 25 percent of his crop to Charley and another 10 percent to Frances. "We probably have lost very near half our fruit now," Hall said. The only reason Jeanne didn't take more fruit is that so much had been gone when it arrived, the growers said.

Even those grim numbers don't tell the whole story. Fruit continues to drop from the trees for as long as two weeks after a major hurricane. And the fruit that survives to harvest will have substantial damage to the peel, which will make it difficult to sell on the fresh market.

Annually about 70 percent of Florida tangerines and about 40 percent of its grapefruit are sold as fresh. But wind-scarring and other hurricane related damage could reduce those percentages by as much as 20 percent to 30 percent in the 2004-05 season, Counter said. That means further economic losses to Polk's growers on top of the crop losses because fresh citrus generally returns more money to them.

Polk is the state's leading producer of tangerines and other specialty citrus. Annually the county produces about 20 percent of specialty citrus. Polk's grapefruit losses could mean a fatal blow to that segment of the industry. The county is the largest producer of grapefruit outside the Indian River growing region, which accounts for about two-thirds of the state's grapefruit annually.

Frances destroyed at least 70 percent of the Indian River crop, by many industry estimates, and Jeanne may have wiped it out entirely. Stan Carter, an Indian River grower with more than 1,000 grapefruit acres, said he did not have even 5 percent of his crop left after Jeanne. Maybe one-third of the state's grapefruit crop remains on the trees, said Arthur "Duke" Chadwell, the manager of the Citrus Administrative Committee in Lakeland, a federal agency that regulates fresh citrus marketing.

Observations from Peter Spyke

The chapter on Hurricane Frances included a series of e-mails sent to clients, friends and family by Peter Spyke, General Manager of Arapaho Citrus Management. Mr. Spyke is a third generation Indian River citrus grower and has been active in the Florida citrus industry for many years. When Hurricane Jeanne threatened the Florida east coast he resumed sending e-mails regarding the effect of the storm.

Hurricane Jeanne—Saturday September 25, 2004

We're down at my Mom's in Davie watching in disbelief as Hurricane Jeanne is about to come on shore at precisely the same point as Frances. The path of the north eyewall—the most intensive part of the storm—will pass directly over our condo and farm again. The path will not be more than a couple of miles off from that of Frances. It's probably as close to impossible as anything could ever be.

There is a big difference between the two storms. Jeanne is traveling 13 mph instead of 3, and Jeanne's sustained winds are about 125 mph instead of 105. There won't be nearly as much rain out of Jeanne, but the wind damage, and storm surge, will be more severe.

I had worked hard Wednesday and Thursday counting fruit with the crop insurance adjuster, and riding groves with fruit buyers, and I had just about come to the conclusion that we were going to come out much better than expected. A lot of grapefruit had been blown off by Frances, but the grapefruit left on the trees was the "inside fruit" which grows within the canopy of the tree. The inside fruit is usually larger in size, flatter in shape, and sweeter than the fruit on the outside of the tree. The price was going to be very high, so between the crop insurance and the higher price for what was left, we stood a good chance of coming out of the whole thing in remarkably good financial condition. Plus, buyers could no longer take grapefruit for granted—they had to pay more attention than usual since the supply would be short, and the fruit they would have received would have been, much better than usual because it would have been inside fruit. There would have been some really spectacular cartons of grapefruit in the marketplace. Grapefruit stood a good chance of actually gaining recognition, or notoriety, in the long run.

Now, of course, we'll have to see. The winds will be out of the same directions, so it's possible that some of the fruit that made it through Frances will survive Jeanne as well. We need to have some left, or else the potential positive outcome may be compromised.

We're of course worried about our home and office in St. Lucie County. The storm surge in Frances came within 4 inches of flooding our condo. The storm surge in Jeanne is supposed to be 4-6 feet higher than it was in Frances. It will be low tide when the eye approaches, so that may help, but we're still concerned.

The grove house lost quite a few shingles in Frances. As a result, we've had a couple of minor leaks in hard rains, but the roof remained fairly water tight in spite of the loss of the shingles. The higher winds of Jeanne, though, may blow off some more or even work on the tarpaper underneath. We may end up joining the blue tarp roof crowd. (Blue tarps to cover roof damage were provided by FEMA and installed by the Army Corps of Engineers.)

Preparation went a lot more smoothly this time around. Most of the plywood was still up at the grove and condo, so we only had to put up a few more pieces we had obtained since Frances, and replace the pieces we took down to let in light. Since we'd had a lot of practice putting up plywood, things went a lot faster this time with fewer expletives. Once again, Cindy and I packed our best clothes and meaningful belongings from the condo into the two 4-runners. Some things were still packed, and we had actually left our photo albums and musical instruments down at my Mom's after Frances. We ran the Arapaho office up till 6:00 yesterday, and still had everything packed up by midnight. The wind woke us up this morning about 5:00, so we got off to a good start.

It's interesting to have to go through this all again so soon. The experience was fresh in everyone's mind, and the shortfalls in preparation were more easily avoided. During the days after Frances, we found that hurricane preparation procedures recommended by the authorities did not include everything we needed to adequately cope with the storm and it's aftermath. We decided that notable omissions from the list of emergency items to stock up on prior to evacuation were beer and anything chocolate.

After it gets done hammering Ft. Pierce and all the groves in that area, Jeanne is supposed to turn north and pass directly over The Orange Shop in Citra. If the eye stays east of the Shop, the fruit should be OK because there are tall oaks on the north and east side. If it moves west of the Shop, we'll probably lose a lot of that fruit, too. That's our Holiday packing fruit, and at this point it's really nice. I also bought a crop of Navels up in Weirsdale since the ones in Ft. Pierce were banged up and blown off. Up until Jeanne I was feeling pretty good about that decision. If they get hammered, I'm not sure where to look next. We're running out of unscathed groves. There are 750,000 acres of citrus in Florida, and by last count about 540,000 acres had been damaged in some way by Charley and Frances.

I'm pretty well in shock. It's hard to maintain my composure, but I'm trying to be careful to not get ahead of things. There is a curfew in place in St. Lucie County until 1:00 pm tomorrow, Sunday. I'm not sure whether we'll head back tomorrow, or wait until Monday morning. I suppose it will depend on whether we lose power here. If we do, we might as well be in Ft. Pierce running the generator. All the Arapaho crew is hunkered down up there, so hopefully they will be OK. Guess we'll have to see.

Tuesday, September 28, 2004

To pick up where we left off—we last wrote from down at my Mom's in Davie (near Ft. Lauderdale) as Jeanne was coming ashore at Ft. Pierce around midnight. I went to bed about that time, and got up Sunday and prepared to make our way back home. We finally got off from Mom's about 11:00 am with the two 4-Runners that contained me and Cindy, most of our important belongings, and our two bird dogs (one of our dogs was really old and didn't survive the trauma of evacuating for Frances, so we lost her before Jeanne). As we drove north on the Turnpike, traffic was mercifully light. At noon on Sunday the storm was still over the state, and we drove through some rain bands although we had waited until the worst of it cleared Ft. Pierce according to the radar before departing. As we neared West Palm Beach, we could see that Jeanne had added another layer of damage over the top of that caused by Frances. Rainfall amounts apparently increased as we moved north. A lot of blue tarps were blown off roofs, and more roofs were damaged.

We arrived at the grove house (the Arapaho office) and could see that we had been whacked again. We'd lost more shingles off the roof, and this time the leaks were significant. We fired up the generator and mopped up the water on the floor and counter tops. A lot of water had run down through the heart cypress ceilings, and a few of the tongue and groove boards have cupped. I'm sure the fiberglass insulation is soaked. Hopefully, the attic fan that pulls air from inside the house up into the attic will help dry it all out before the mold sets in. The kitchen also had a lot of water. The roof is clearly compromised, and we will have to move that up on our priority list of things to do.

We lost one side of our shop, which is made of tin, and everything inside the barn and shop was soaked. The gutter blew off of Brenda's trailer porch, and a few of the aluminum sections of the porch peeled back. A section of the roof of a storage building was also peeled back. Otherwise, everything else came through OK, including our flats boat that we had barricaded in with trucks and tractors.

Robert had switched off the electricity to the Shadow Lakes pumps Saturday afternoon at 4:00, and hooked up the tractors to the pumps and got them going before a curfew went into effect. Since the storm was moving so fast, the tractors could run long enough on one tank of fuel to make it through the worst part, so no one had to ride it out in the grove to keep them fueled like Danny had to do in Frances. Bobby got the pumps going in Ft. Pierce Sunday morning about 8:00—there is more water in the Ft. Pierce groves than I have ever seen. It's running over the tops of the control structures in the North St. Lucie Water Control District. All three of my groves in the area were completely under water—tree trunk to tree trunk—with water running through them out of the canals since it had topped the roadways that border the canals. This was the third time they flooded in three weeks, and this time was worse than the first two events. More than likely I will just write off the Rock Bottom grove completely since it was the worst, and try to collect tree insurance and replant. The Home Grove and 36 have a little more control, so we're trying to pump them out in time to save the trees. Even so, the roots have undoubtedly been damaged, so there will be an effect for a year or two.

I have found that, despite the importance of the information, my first reaction to both storms has not been to drive out in the groves. For one thing, they're flooded, so it's just a mess that's hard on wheel bearings and u-joints in the vehicles, and it tears up the roads and beds. Whatever fruit has fallen is just the first wave—more will drop for weeks as injured fruit is shed, and broken stems let go. So, other than getting an initial idea, there's not much to do. The first priority in the groves is to get rid of the water, and that will occupy most people in the company for the next week. Overwhelmingly, though, our first efforts after the storm are related to people—getting in contact, ascertaining their situation, and organizing both a work plan and relief effort as necessary.

Labor was tight before Jeanne since any available picker or grove worker is picking up debris at wages significantly higher than we pay. We had sort of reached a workable solution with our permanent people and a contract crew. The highest priority is that one of our customers had 12,000 young citrus trees blown over, and a bunch of newly-planted palms had blown out of the ground and had to be re-planted. We had just finished that work after Frances, and had actually started planting trees again. The winds from Jeanne were higher, and the ground was totally saturated, so every darned one of them, including the ones we just planted, blew over again. We were holding the line on labor costs since we were steady customers of the contract crews, but I bet that loyalty will be sorely tested when we try to get them to start all over for the same price.

The citrus packinghouses in the area were decimated. A lot of them had some minor roof and siding damage from Frances that hadn't been repaired. These openings allowed the wind to get inside during Jeanne, and they are a lot worse off now. St. Lucie Packing, the one remaining packinghouse out of 4 that can pack fresh lemons that was operable after Frances, was our last hope to get our lemon crop packed IF we could find pickers. I went by there today and the roof and one side of the house are very badly damaged. They're responding quickly—pulling all their field crews in to see if they can repair the damage enough to run. We actually started running their pumps and screw gates in their groves near our groves so their people could go work on the packinghouse. There are many more packinghouses that are in worse shape—many may not be able to run this year at all. We were wondering how they were going to deal with the shortage of grapefruit—there wasn't enough to go around—and Jeanne may have decided that to some degree.

It's now Tuesday morning as I'm finishing this e-mail. The generator conked out last night, so we lost the comfort level it provides, and we weren't able to open the office this morning since the computers and phones won't work without power. We had a fairly complete crew show up today, including the mechanics, so while they worked on the generator we got people lined out for work. The generator was initially pronounced DOA, so we hooked up James' gas welder out of his truck and got that going. It doesn't have a 240 outlet, so we wired together the two 120 outlets and it will run the house, but it doesn't put out enough amps to start the larger motors like the fridge. The battery backups are beeping incessantly probably due to power sags. In the meantime, James took a condenser out of a Chevy truck and put it in the Briggs and Station engine on the generator and got that going. So, we'll re-wire back to the generator and will hopefully be OK.

I honestly don't know what kind of experience Cindy and I would have had in these two storms were it not for the wonderful, resourceful people who work for us at Arapaho and The Orange Shop. They have done an amazing job of making it happen under very challenging circumstances. We are blessed.

The weather has actually been beautiful. The nice thing about a storm going north of us is that it pulls in some dry air from the Midwest on the west side and wraps it around to Florida. This helps things dry out, and is also not as oppressive when we're trying to live without AC. Last night was a cloudless, full moon. The only disruption was from pulling the rope on the generator. Fortunately, the artesian well that supplies water to the house puts out enough pressure to get a trickle of water out of a shower head, so we were able to take a long, leisurely, cold shower.

All this is wearing on everyone. I haven't had a haircut in 5 weeks, and for most of the last month we've been living out of the back of our 4-runners. Just getting through a day takes a lot of effort, so not much forward progress is possible. Needless to say, we're hoping that this is the last storm. I had the thought that it would be almost impossible for another storm to hit us, but then realized that it was almost impossible for two storms to hit us in precisely the same point, and remembered that Polk County in central Florida has had all three eyes pass over it from Charley, Frances, and Jeanne. This is apparently not the time to make a trip to Las Vegas to play Blackjack—our luck is running bad at the moment.

I'm off to unload the office from the red 4-runner and hook up the server and computers. We need to run payroll tomorrow, and it's time to close out our billing for the month. September is normally a pretty good month for Arapaho—we'll see how it turned out this year.

Saturday, October 2, 2004

Wes Brumback had sent down his generator with Jessica on Wednesday, which was a good thing because our old faithful Homelite finally died Wednesday night. We were able to switch over to Wes' and keep going. His was great—it has a huge gas tank and will run all night without refueling. A real luxury.

We opened the office on Tuesday, but really didn't get settled until Thursday. It took that long to get all the electronics and ventilation sorted out. The generators don't put out steady enough power to make the battery backups happy—the backups kept chirping as the power sagged. We kept switching around the backups on different house circuits, and turning lights on and off until we reached an equilibrium that allowed us to run the server, two desktops, two desk fans and the attic fan without incessant beeping. We also had to dig around in the boxes we evacuated to find the checks, invoices, and so forth. We can work until about 1:00 in the afternoon unless we get some clouds or rain before it gets too hot. There was power back across the street last night, so hopefully we'll get ours back by Monday.

The groves are a wet, stinking mess. Rotten fruit from Frances is floating around with fruit blown off by Jeanne. Again the fruit plugged up the drain pipes and the guys had to get down in the gook in water up to their chests to clear the pipes so water could drain. Nobody would do it voluntarily, so Richard had to go out with one or two guys and get down in it with them so they wouldn't balk. Just another one of the true Arapaho and Orange Shop heros in this deal. We have more crews out standing up the young trees we stood up after Frances blew them down—now we're doing it all again. We're still finding labor at below debris-pickup prices, so at least we're holding the cost down for our customers. We haven't tried to do much tractor work in the groves because they're still wet, plus I don't have a good plan for moving forward. So, the hand labor and tractors on pumps will be our main source of revenue for the entire month of September. Thank goodness for both us and our customers that we were in a position to provide those services, though—we saved a lot of trees from serious damage.

The insurance adjusters have come to the Arapaho grove house and Cindy's duplex. Both will get new roofs. Fortunately, we have stopped the leaks, so we can wait a bit before hiring someone to put on the roof. The adjuster that inspected the grove house said that roofers were charging $250 per 100 square feet now, while 6 months ago it was $135. He suggested that if we wait a few months it will go back down. Maybe—there are a lot of roofs to fix. I may see if we can figure out how to do the Arapaho roof ourselves. Cindy's is a flat roof, and you really need to be set up for that, so she'll need a pro to fix the roof on the duplex.

The adjuster strongly recommended that we spend the extra money for "architectural shingles" instead of going back with regular "three tab" shingles. After inspecting homes in the paths of all 4 hurricanes (Charley, Frances, Ivan, and Jeanne), he said that there was no comparison between the two. He even said that three tabs may be outlawed before much longer in Florida since there was so much damage. We've noticed the same thing as we have looked around.

We lost quite a bit more fruit in Jeanne. Grapefruit and Navels, especially, were stripped of more fruit, and we can expect it to keep falling for the next 3-4 weeks. All our work on crop insurance claims from Frances is out the window—we have to start over. At this point, there are no accurate estimates of loss. Between the two storms, I would guess that 75% of the grapefruit statewide has been lost—probably 80-85% in the Indian River. Jeanne blew off more oranges and tangerines than Frances, effectively answering the question as to whether it's more destructive to have higher winds for a short period of time vs. less wind for a long period of time. The higher winds in Jeanne blew off more fruit and caused more damage to trees and buildings. No question about it.

Fortunately, we will still have enough fruit for The Orange Shop, so this is one year in which being small is a blessing. I am so happy that I called the Monday after Frances and bought a crop of Navels up in Weirsdale. We've lost most of them here, and Weirsdale shouldn't have lost more than 30-40%. We'll have enough grapefruit and other varieties here to supply the Shop and probably two of the other gift fruit shippers that I work with. "I'm very glad that we have fruit for the Shop—we are looking forward to a good season up there if the tourists come down in usual numbers. Also, cash inflows from crop insurance and government disaster assistance, along with fruit income, should be sufficient for our grove care customers to take care of their trees to produce a crop for next year. So, both Arapaho and The Orange Shop should be able to hang in there. Even with reduced budgets, we should be able to adjust and deal with it.

The big commercial packinghouses and harvesting companies are in trouble, though. In many cases, their annual revenue will be a goose egg, and they have to fix the packinghouses to boot. I have many good friends who are in trouble because of this.

Jeanne hit Indian River County (Vero Beach) hard. They had come through Frances better than St. Lucie (Ft. Pierce) or Martin (Stuart), so there was still a lot of fruit left for the packinghouses to run. Jeanne came through and wiped out Indian River County and most of St. Lucie County, and did a number on Martin County as well. There's simply not very much fresh fruit available. Buyers have spanned out across the rest of the state trying to find crops to run. A lot of the packinghouses sustained major damage from Jeanne and will not try to run at all—there's not enough fruit and not enough time to undertake major repairs before the season begins. In fact, the season has actually begun—there are a few houses in central Florida that started running fruit last week before Jeanne, and are picking again this week.

That's going to cause a major hit to our local economies. The income from the fruit itself, which supports growers and grove caretakers, is only about ⅓ of the value of packed fruit ready for market. Harvesting and hauling is usually another ⅓, and packing is the rest. The growers will get some income from crop insurance and government disaster aid to help close the gap for their ⅓, but there will probably not be disaster aid for the packinghouses or harvesting companies. That means that as much as ⅔ of the economic impact of this year's citrus crop on our local economies will be lost.

We had originally planned to open The Orange Shop on October 8th—next Friday. After Frances, we moved it back to the 15th. I'll be going up next week to see if that's still a possibility. The Shop didn't receive much damage—we lost a couple sections of the awnings—but Jeanne knocked the power out and blew trees down on homes in the area, so we lost some time there while people were dealing with all that. Which is OK—the number one priority is people in this deal. We'll try to get our focus back next week and see what we can do. Becky has already started getting ready to open—the floor waxing people will be in Monday and she will start ordering merchandise to stock the shelves. I'll be finalizing the fruit choices for opening, lining up the pickers, servicing the trucks, and tweaking the packinghouse to start running fruit. We have a lot to do in two weeks.

It occurred to me that for some period of time, we will simply have to get used to operating in a state of disorganization. Everything we own has been turned upside down. Clothes are packed in duffel bags, furniture and belongings have been moved around between three houses—my Mom's (we left some things there), the grove, and the condo. The office has stuff strewn all about from the boarding-up and evacuation processes, and from the generator-induced relocations. It would be nice if we could just take a break and get organized. Unfortunately, that's not an option. We have to make some very right moves immediately in the groves to minimize tree damage, fruit loss, and financial impact. We have to open The Orange Shop and finish up the things we had planned to do up there before we opened while we are operating. I don't see how we can carefully plan all of this—it will be hip-shooting at least until January as near as I can tell, since Thanksgiving will be here in about two minutes.

Fortunately, we have the right bunch of people to do this—they are all seasoned pros with good judgment. Things will either feel right or they won't—and that will be the decision criteria rather than spreadsheets and calendars. I believe that we can do it, and do it well. That's the plan—rely on our people, work hard, and have faith that the answers and solutions will come when we need them.

I suppose that this message sort of closes the loop. We've worked through two hits and two threats from hurricanes in the last 7 weeks, and now we're back home with all our stuff more or less intact. Clearly, nothing like this has ever happened. The last time storms hit with this frequency, hardly anybody lived in Florida—too much water and too many skeeters and not all that many people in the country. Now there are millions of people living under three-tab shingles near the coasts. Hopefully, this is really the end of the threats and hits for this year. Actually, forever would be OK with me.

Wednesday, October 20, 2004—The Aftermath

It's now about a month since Hurricane Jeanne came through. Life is beginning to resemble what it was pre-hurricane, but it's still a long way from normal.

The clean-up job is still very much under way. The trees and brush have largely been removed, but the sides of the roads are still littered with tarpaper, drywall, furniture, and other detritus from damaged buildings. A significant percentage of dwellings are uninhabitable. I stopped by a friend's condo today that had roof damage in both storms. The unit had to be completely gutted since the ceilings collapsed, which was a blessing since the blown-in insulation was completely moldy. It's now only concrete floors, walls, and studs, and will have to be completely rebuilt.

Most roofs have been either covered with blue tarps or patched. We got our first estimate for fixing the office roof, and the cost is about 3 times what it would normally be. Fortunately, it's not leaking so we'll wait a bit until the craziness diminishes a bit. We received a check from the insurance company, but since the roof was damaged in both storms, they took out two deductibles, so there isn't enough to cover the cost. It'll certainly help, though.

The local hotels and motels are completely booked by displaced storm victims. FEMA is helping cover the costs for a period of time. It will take longer than the time period for repairs to be completed, though. I imagine that it's worse here and over in Punta Gorda on the west side of the state.

The schools are trying to figure out how to make up for the lost days. They're talking weekends and holidays. The state of Florida has a testing program that measures the effectiveness of the educational programs, and they're talking about making this area exempt because the kids lost so much time.

Business-wise, we're pretty fortunate, but we will feel the effects. The fruit that remains on the trees down here is scarred, and the maturity development has just stopped. The trees lost so many leaves that they can't produce sugar as fast as usual. Navels are particularly affected. In most cases, the sugar levels in the fruit (°Brix) are actually lower than they were on August 19, instead of higher.

The price of fruit has skyrocketed. The growers in California and Texas have already started receiving calls from overseas buyers looking for product. Grapefruit, especially, will be in short supply. I suspect that the buyers will have to access crops in Cuba (if they have any left after Charley), Honduras, Argentina, South Africa, and Australia in addition to the US. My friends in Turkey should have a very good year with their grapefruit, although their crop is down a bit due to natural cycles. The orange juice supply won't be affected as much since Brazil already has well-established transportation and infrastructure to move juice up to the U.S. Tangerines and other specialty varieties weren't hit as hard as grapefruit and Navels.

We started picking last week, and opened The Orange Shop on the 15th. It was a scramble, but we made it. The first few days were about 30% off from last year. The fruit doesn't taste all that great, and I don't think we have as much out of state traffic. Probably a couple of reasons for the latter—the lack of lodging may have something to do with it, the publicity may have caused people to cancel plans, and of course gas is expensive. Hopefully it will pick up as we approach the traditional snowbird season.

We bought another Navel crop in Weirsdale to share with one of my friends who is a gift fruit shipper, and those Navels are fine. We paid 4 times what Navels sold for last year, but we danced a jig when we made the deal because we have enough Navels to supply our gift fruit buyers. The internal quality is actually better than in Ft. Pierce since the trees didn't lose any leaves that far north. The problem is they're still green, and we don't have degreening facilities at the Shop. Since we can degreen at a commercial packinghouse in Ft. Pierce, we'll stay with the fruit down here until the Weirsdale Navels break color. There was a good cold front last weekend, so they should start breaking very soon. It takes cold weather to make the oranges turn orange—the cold weather causes the trees to release ethylene, which denatures chlorophyll. The oranges and yellows are in the peel all along, but the chlorophyll is the strongest pigment. When it fades, the other colors show through. All we do in degreening rooms is to circulate the air with fans and inject about 2 ppm of ethylene into the air flow, which mimics the natural process. It's normal this time of year for the fruit to be ripe, but still green, so all we do is help nature out.

FEMA has offered a program for people who have lost work due to the storms. They'll pay for up to 26 weeks, at which point unemployment can kick in if there are still no jobs. I had a negotiation with our pickers today because they're signed up, and are nervous about having enough fruit to pick to come out ahead if they go off the government aid. Other people are having a problem because their crews are drawing FEMA or unemployment. I'm trying to devise a way for the pickers to come out better by coming back to work, but it's tough to come up with something that works. They're understandably worried because the crop is so short that there's no way they'll make enough picking to earn a living through the entire year unless we do things differently. I'm hoping that we'll be able to put together enough picking to keep them steadily busy in order to make it worth their while to come back on the payroll. It looks like we'll be more heavily involved with pickers and picking than normal. One more layer of complexity required to get things done that wouldn't be there without the storms.

I called the Department of Plant Industry this week to check in with them. Their headquarters are in Winter Haven, where Charley caused a lot of damage prior to being hit by both Frances and Jeanne. We had submitted budwood from a Navel grove that we used to manage to them so that they could test it for viruses. "Clean" budwood is a big deal for us—we try to use propagating material that doesn't have diseases that will shorten the life of the trees. If the budwood has viruses, the DPI can "shoot-tip graft" new trees and clean up the material. The viruses are not present in the few cells at the growing tips of the shoots, called meristems. The technicians can use a microscope to remove these few cells and grow them out in a nutrient culture, and then eventually recreate the parent tree sans viruses. Once the propagating material is cleaned up, we grow out the trees and use them as a source of budwood for the commercial trees in the groves. The entire process takes about 5-6 years to complete.

One of the most damaging viruses, Tristeza (Spanish word for "sadness"), is spread by aphids as they feed on infected then healthy trees. In order to maintain Tristeza-free budwood, the DPI maintains a collection of clean source trees in screen houses that keep the aphids out. They test the trees routinely, and the method works.

Unfortunately, all of the screen houses were totally destroyed. This essentially means that the source budwood has been compromised, because the aphids have had free access and the trees growing outside the screen houses had Tristeza. So, the DPI will have to start all over with empty screen houses, and populate them with trees only after they've been shoot-tip grafted. There are hundreds of cultivars of citrus grown in the state, so it will be years before the collection is re-established. There are some other screen houses in Immokalee with a budwood collection, but that area has canker nearby, so there's apparently some reluctance to bring up budwood to Winter Haven. Just a note of trivia that might be of interest to you non-citrus people out there. The impacts of the storms run far deeper than the more obvious property damage issues.

It looks like the citrus growers are going to get some financial help from the government. Basically, smaller operations will benefit the most—there are limitations that essentially reduce the amount of help larger operations can expect. That will be a godsend, because it will provide financing to help grow another crop, and will help keep our local economy going. Even with the help, it will be a very lean year for the growers. It will even be a leaner year for the packinghouses—there's not really any help for them.

Everyone is clinging to normality wherever we can find it. I spent most of the last week up at The Orange Shop, helping to get it opened up, and Cindy came up for the weekend with some friends. We went to the Gator game and watched them beat Middle Tennessee State 52-16, so that was fun. It felt strange to do the normal game things like cheering and eating bratwurst sandwiches—it almost seemed surreal after all the hullabaloo. Some of the Shop people came with us, and it reminded me that people pull closer in these kinds of situations, so a good football game is almost therapeutic (especially if our team won). I'm sitting here watching the Yankees and Red Sox, game 7, and it looks like the Sox have pulled off the impossible unless the Yanks come back in the 9th. These simple pleasures have never seemed so important.

Bobby finally took down the plywood at the office and pronounced that we aren't going to get any more storms. In retrospect, that was a very positive move. He's leading the way of letting go, which is what we all need to do. We'll be playing catch-up for awhile yet, though We're behind on our grove work, and evaluating the things that need to be done and those that can be skipped to save money and time. We have a couple of irrigation systems we need to get installed by the end of the year, and the pests that took a hiatus during the storms are starting to build again. Picking has begun, so we have to coordinate all that.

Time is now the critical factor that has replaced generators, gas, ice, and food supplies that were our major focus only a few days ago. I've stopped attending any non-critical meetings, or even thinking about long-term issues. We won't be able to recover the lost time, but we should be able to work through things well enough to get back ahead far enough to regain control over a lot of the more routine issues. Realistically, I expect that to take until about April or May. Until then, we'll have to hustle every day.

The Citizen Weather Observer Program (CWOP)

Citizen Weather Observer Program (CWOP) stations made important contributions to data collection during Hurricane Jeanne. According to Russ Chadwick, Patricia Miller and David Helms of NOAA in *Hurricane Jeanne Landfall Observed by Citizen Weather Observer Program Stations*, the 81 mph gust recorded by CWOP station CWO262 in Port St. Lucie was one of the highest gusts recorded by any Florida station before being driven off the air by the western eyewall of Jeanne. However, despite two brief outages, CWOP station K4ZZR, 3.4 miles east continued to report, and measured the lowest pressure observed in the United States for Jeanne at 954 mb.

During the period when Jeanne was over the state of Florida, the above two CWOP stations continued to send data from the St. Lucie County area even after up to a million people along the Florida east coast had lost power and all three of the NOAA Automated Surface Observing Systems (ASOS) between Melbourne and Miami had gone off the air.

The Citizen Weather Observer Program is a cooperative venture between the NOAA Forecast Systems Laboratory (FSL) located in Boulder, Colorado, the National Weather Service and individuals with private weather stations and internet access. Many of these participants are licensed amateur radio operators such as WX4NHC, the amateur radio station at the National Hurricane Center in Miami. Other stations are owned by schools and state and federal agencies.

In July 2005, approximately 5,000 individuals belonged to the CWOP program, but it was estimated that up to 10,000 would be members by 2006. Individuals with home weather stations and internet access are invited to join the program.

The Impact of the 2004 Hurricanes on Fresh Fruit Shipping

According to the Fresh Shippers Reports from the Citrus Administrative Committee, the total shipments of fresh fruit measured in 4/5 bushel boxes, dropped from 52,057,489 in the 2003-04 season to only 30,194,004 in the 2004-05 season, a decrease of 21,863,485 or 42 percent.

Grapefruit suffered the most severe losses as a result of the damage by Hurricane Frances and Jeanne to the grapefruit groves in the Indian River area. Grapefruit shipments in 4/5 bushel boxes dropped from 30,350,424 during the 2004-05 season, a loss of 16,940,934 boxes or 56 percent. Orange shipments dropped from 12,310,291 to 9,809,929 boxes, a 20 percent loss. Early tangerines fell from 4,230,386 to 3,262,338 boxes, a 23 percent loss. And Honey tangerines from 3,968,118 to 2,771,175 boxes, a 30 percent loss.

In 2003-04, the top 24 packers each shipped over one million boxes, but in 2004-05 only 11 packers reached the one million box level. Tables 4-3 and 4-4 show the numbers of boxes packed by each of the top 20 packinghouses in each of the two seasons. Note that in 2003-04 grapefruit represented 58 percent of the shipments and oranges 24 percent, but in 2004-05, grapefruit had dropped to 44 percent and oranges had risen to 33 percent.

TABLE 4-3. SHIPMENTS BY THE TOP 20 PACKERS IN THE 2003-04 SEASON.

	Shipper	Total 4/5 Bushel Boxes Shipped
1.	Greene River Packing	3,612,626
2	Dundee CGA	3,603,386
3.	Bernard Egan Company	2,963,651
4.	Heller Brothers Packing Corp.	2,368,300
5.	Premier Citrus Packers LLC	1,800,560
6.	Lake Placid Groves LLC	1,779,722
7.	A. Duda and Sons, Inc.	1,776,815
8.	Haines City CGA	1,539,619
9.	IMG Citrus, Inc.	1,532,352
10.	Ben Hill Griffin, Inc.	1,508,687
11.	W. G. Roe & Sons, Inc.	1,420,765
12.	Packers of Indian River	1,411,614
13.	Indian River Exchange Packers	1,245,271
14.	Riverfront Packing Co. LLC	1,164,610
15.	United Indian River Packers	1,161,155
16.	Emerald Packing Co., Inc.	1,155,183
17.	East Coast Packers, Inc.	1,148,341
18.	Graves Brothers Company	1,146,176
19.	Leroy Smith, Inc.	1,124,207
20.	Hunt Brothers Cooperative	1,069,413

Source: Fresh Shippers Reports. Citrus Administrative Committee, Lakeland, FL 2006.

TABLE 4-4. SHIPMENTS BY THE TOP 20 PACKERS IN THE 2004-05 SEASON.

	Shipper	Total 4/5 Bushel Boxes Shipped
1.	Dundee CGA	3,172,562
2.	A. Duda & Sons, Inc.	2,004,287
3.	Bernard Egan Company	1,722,634
4.	Heller Brothers Packing Corp.	1,595,717
5.	Ben Hill Griffin, Inc.	1,549,573
6.	Haines City CGA	1,423,465
7.	Lake Placid Groves LLC	1,362,983
8.	Greene River Packing, Inc.	1,355,116
9.	Packers of Indian River, LTD	1,324,728
10.	Hunt Brothers Cooperative	1,226,746
11.	Emerald Packing Company	1,103,112
12.	W. G. Roe and Sons	877,551
13.	Leroy Smith, Inc.	831,789
14.	Peace River Packing Company	692,766
15.	Premier Citrus Packers, LLC	650,409
16.	Riverfront Packing Company	616,410
17.	Minton Sun, Inc.	587,230
18.	Waverly Growers Coop.	572,904
19.	Hogan and Sons, Inc.	529,129
20.	Rio Citrus	522,356

Source: Fresh Shippers Reports. Citrus Administrative Committee, Lakeland, FL 2006.

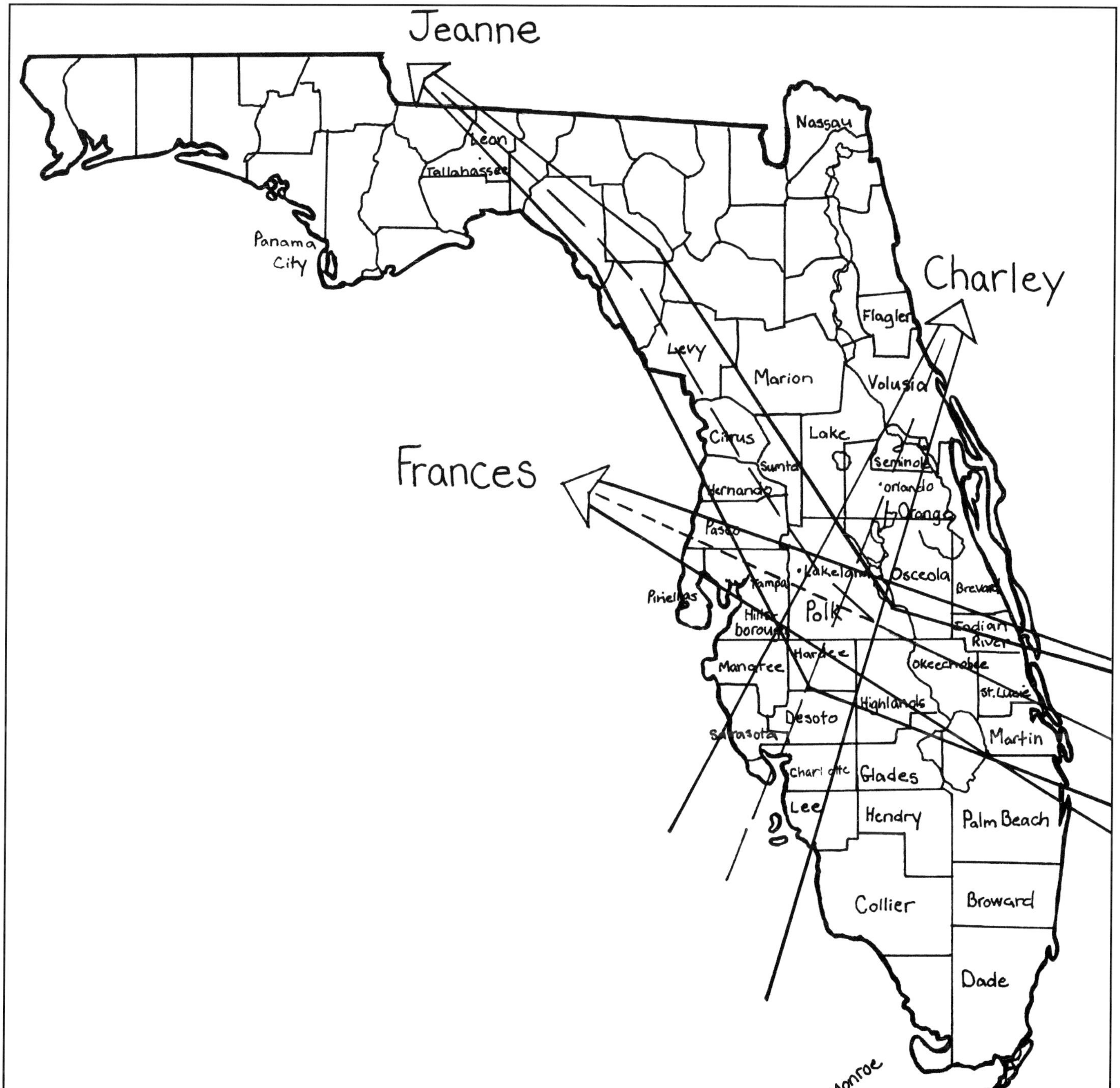

Figure 4-6. The courses of Hurricanes Charley, Frances and Jeanne as they crossed peninsula Florida on August 13, September 6 and September 26, respectively. Polk county was the only county to feel the wrath of all three hurricanes, a historic first. Sketch by Brooke Elizabeth Attaway of Lakeland, FL.

CHAPTER 5

EFFECT OF THE 2004 HURRICANES ON PRODUCTION AND VALUE OF FLORIDA CITRUS

The Numbers

According to the Florida Field Office of the National Agricultural Statistics Service (NASS) in Orlando, the total production of all varieties of Florida citrus for the 2004-05 season was 169.1 million one and three-fifths bushel boxes, the lowest production since the 154.2 million boxes after the catastrophic Christmas freeze (Attaway, 1997) of 1989 (Table 5-1). This was a decrease of 42% from the 291.8 million boxes harvested during the 2003-04 season, and was the first season since 1991-92 that the total production was below 200 million boxes. Production was lower for all varieties except tangelos which were grown in counties which escaped a direct hit by the hurricanes.

The orange crop including all varieties was 149.6 million boxes, the lowest since the 139.8 million boxes in 1991-92. The production of early, midseason and Navel oranges decreased by 37% and Valencia production by 39%. At 2.5 million boxes, the Navel crop was the lowest since the 1935-36 season. Grapefruit production at 12.8 million boxes, down from 40,900 boxes in the 2003-04 season, was also the lowest since 1935-36, and reflects the vulnerability of the heavier fruit to damaging winds. Early tangerines, including the Fallglo and Sunburst varieties, fell from 3.6 million to 2.45 million boxes while honey tangerine production fell from 2.9 million to 2.0 million boxes (Table 5-2). Murcotts and Honey tangerines are two names for the same variety.

TABLE 5-1. FLORIDA CITRUS PRODUCTION OF ALL VARIETIES BY YEAR FROM 1989-90 TO 2004-05.

Season	1,000 1-3/5 bushel boxes
1989-90	154,185
1990-91	205,660
1991-92	191,815
1992-93	251,540
1993-94	235,760
1994-95	271,020
1995-96	265,375
1996-97	295,315
1997-98	304,450
1998-99	243,165
1999-2000	298,430
2000-01	278,290
2001-02	287,030
2002-03	250,850
2003-04	291,800
2004-05	169,100

Source: Florida Agricultural Statistics Service, Citrus Summary 2004-05.

The Value

The $742.2 million preliminary value of the 2004-05 citrus crop, the lowest since 1985-86, was down 16.7 percent from the 2003-04 season's revised value of $891.5 million (Table 5-3).

After Hurricane Frances and Hurricane Jeanne devastated the Indian River District, the price per box for grapefruit was the highest on record for both white and colored varieties. The price for white grapefruit increased from $1.88 to $11.95 per box and colored grapefruit increased from $4.17 to $14.27 per box. Obviously, the grapefruit grower whose crop survived the hurricanes had his most profitable season in many years. In the interior counties, the price per box for Temples increased from 1.07 to 2.48 per box, early tangerines from 6.15 to 10.12 per box and Honey tangerines from 9.16 to 14.36 per box. The tangerine grower whose fruit survived the winds also had a good year (Table 5-4).

Whether a grower was among the lucky few had everything to do with "location, location, location." Those who were fortunate to have groves in Hendry and Collier Counties largely escaped the onslaught of all three hurricanes. Hendry County growers lost 13% of their orange crop and Collier County growers lost only 8%, but the very unlucky growers in both Hardee and DeSoto Counties lost 53% of their oranges (Table 5-5), Hurricane Charley being the culprit. Meanwhile orange growers in St. Lucie and Indian River Counties who hosted both Frances and Jeanne each lost over 60% of their orange crops. Polk County growers who suffered the wrath of all three hurricanes lost 40%, the winds having subsided by the time they reached the ridge.

While the east coast counties lost much of their orange crop, oranges are not really the money crop along the Indian River which is grapefruit, and the combined effect of Hurricanes Frances and Jeanne produced the greatest weather disaster in the history of the Indian River District. Again it was location. St. Lucie, Indian River and Martin counties are relatively warm counties compared to Lake County and the counties along the ridge

TABLE 5-2. DECREASE IN CITRUS PRODUCTION BETWEEN THE 2003-04 AND 2004-05 SEASONS BY VARIETY.

	2003-04	2004-05	% Loss
Variety	1,000 1-3/5 bushel boxes		
Early-Midseason and Navel oranges	126,000	79,100	-37.2
Valencia Oranges	116,000	70,500	-39.2
White Grapefruit[1]	15,900	3,400	-78.6
Colored Grapefruit	25,000	9,400	-62.4
Temples	1,400	650	-53.6
Tangelos[2]	1,000	1,550	+55.0[2]
Early Tangerines[3]	3,600	2,450	-31.9
Honey Tangerines	2,900	2,000	-31.0

[1]Includes seedy grapefruit.
[2]Tangelo production was greater.
[3]Fallglo and Sunburst varieties.
Source: Florida Agricultural Statistics Service, Citrus Summary 2004-05.

as a result they did not feel the major devastation of the 1962, 1983, 1985, 1989 and other severe freezes. For one year in the 1970s, Lake County produced more oranges even than Polk County, but the freezes ended Lake County supremacy for good in the 80s (Attaway, 1997). A major hurricane in 1949 did result in major fruit loss and grove damage on the east coast, but the area had never suffered two major hurricanes which followed the same path in the same year. As a result of Frances and Jeanne, St. Lucie and Indian River Counties lost over 80% of their grapefruit crops and Martin County over 70% (Table 5-6) Polk County, again feeling the lessened winds away from the coast, lost over 60% of its grapefruit crops. While real estate development has affected Brevard and Palm Beach Counties to a great level, and they are no longer major citrus producers, it is significant that both these counties lost over 80% of their remaining grapefruit crops.

Tangerine production prior to the 2004 hurricanes was greatest in Polk County with 1.46 million boxes. Lake County grew 813 thousand boxes, St. Lucie County 463 thousand boxes and Palm Beach County 382 thousand boxes. Percent losses in these counties were Polk-31%, Lake-35%, Highlands-31%, St. Lucie-68%, and Palm Beach-30% (Table 5-7). The direct hits by Frances and Jeanne did the greatest damage to St. Lucie County coastal groves. Surprisingly, Hendry County increased production.

Six counties, Hardee, Hendry, Highlands, Lake, Polk and St. Lucie, are the most important for specialty fruit, mostly Temples and tangelos. Again St. Lucie County was devastated by the loss of 83% of its specialty fruit, while Lake County actually produced 89% more specialty fruit in 2004-05 then it did in 2003-04 (Table 5-8), and Polk County 26% more. The increase in Lake County was chiefly attributed to tangelos.

Production by Area

NASS reports that production in 2004-05 was down in all Florida commercial citrus production areas from the 2003-04 season. The Southern area was down the least at 14 percent, while the Indian River area was down the most at 76 percent. Two hurricanes passed directly over the Indian River area causing major reduction of the citrus crop. The Western area, also severely affected by a hurricane's direct hit, was down 52 percent. Production in the Central area was down 41 percent, while the Northern area was down 36 percent.

Post-hurricane citrus production decreased in each of the 30 citrus growing counties. Hendry County led in total production with 29.6 million boxes, followed by Polk (24.8 million boxes), Highlands (21.3 million boxes), DeSoto (13.6 million boxes) and Collier (10.5 million boxes). These top five counties produced 59 percent of the citrus crop.

TABLE 5-3. LOSS OF ON-TREE VALUE OF ALL CITRUS VARIETIES FROM 2003-04 TO 2004-05.

Crop year	Production in millions of 1-3/5 bushel field boxes	Value of production in millions of dollars
2003-04	291,800	891.5
2004-05	169,050	742.2

Source: Florida Agricultural Statistics Service, Citrus Summary 2004-05.

TABLE 5-4. INCREASE OF ON-TREE VALUE PER BOX BY VARIETY FROM 2003-04 TO 2004-05.

Fruit Type	Crop Year	Production (1,000 1-3/5 bushel boxes)	Price per Box ($)	Value of Production (million dollars)
All Oranges	2003-04	232,107	2.89	699.927
	2004-05	149,600	3.40	508.548
All Grapefruit	2003-04	40,900	3.28	134.123
	2004-05	12,800	13.65	174.141
Temples	2003-04	1,400	1.07	1.502
	2004-05	650	2.48	1.615
Tangelos[1]	2003-04	1,000	7.48	7.484
	2004-05	1,550	2.45	3.794
All Tangerines	2003-04	6,500	7.46	48.464
	2004-05	4,450	12.02	53.503

[1]The price per box for Tangelos was less due to a larger crop.
Source: Florida Agricultural Statistics Service, Citrus Summary 2004-05.

In 2004-05 Hendry County produced the most oranges followed by Polk, Highlands, DeSoto, and Hardee counties. St. Lucie (2.6 million boxes) remained the top grapefruit producer followed by Hendry (2.3 million boxes) and Indian River County (2.0 million boxes). Normally, St. Lucie and Indian River counties produce about two-thirds of the state's grapefruit; however, they produced only 36 percent of the state's grapefruit crop in 2004-05. Colored varieties have dominated grapefruit production for the past 15 seasons and now comprise 73 percent of the crop.

The majority of the specialty crops, tangerines, Temples and Tangelos are concentrated in the ridge counties of Polk, Lake and Highlands. Polk County leads in production of Tangelos and tangerines, while Highlands County was the top producer of Temples.

Estimates of county production were prepared from objective survey data used in forecasting citrus crop production. The apportionment of final harvest to the counties was based on bearing trees from the biennial Commercial Citrus Inventory; the limb count survey,

TABLE 5-5. DECREASE IN ORANGE PRODUCTION (ALL VARIETIES) BETWEEN THE 2003-04 AND 2004-05 SEASONS BY COUNTY.

County	2003-04	2004-05	Decrease	% Loss
		1,000 1-3/5 bushel boxes		
Brevard	1,288	564	724	–56.2
Charlotte	5,589	4,646	943	–17.8
Collier	10,601	9,707	894	– 8.4
DeSoto	28,418	13,328	15,090	–53.1
Glades	3,646	3,346	300	– 8.2
Hardee	21,567	9,975	11,592	–53.7
Hendry	30,558	26,587	3,971	–13.0
Hernando	369	207	160	–43.3
Highlands	32,916	20,076	12,840	–39.0
Hillsborough	7,578	3,732	3,846	–50.7
Indian River	5,715	2,128	3,587	–62.7
Lake	5,863	3,550	2,313	–39.5
Lee	3,096	2,494	602	–19.4
Manatee	8,838	4,562	4,276	–48.4
Marion	367	208	159	–43.3
Martin	11,292	7,926	3,366	–29.8
Okeechobee	3,440	2,981	459	–13.3
Orange	2,025	1,242	283	–14.0
Osceola	5,576	3,306	2,270	–40.7
Palm Beach	890	339	551	–61.9
Pasco	4,220	2,579	1,641	–38.9
Polk	37,174	21,998	15,176	–40.8
St. Lucie	9,744	3,496	6,248	–64.1
Sarasota	393	200	193	–49.1
Seminole	386	210	176	–45.5
Volusia	115	140	+ 37	+32.2

Source: Florida Agricultural Statistics Service, Citrus Summary 2003-04 and Citrus Summary 2004-05.

TABLE 5-6. DECREASE IN GRAPEFRUIT PRODUCTION BETWEEN THE 2003-04 AND 2004-05 SEASONS BY COUNTY.

County	2003-04	2004-05	Decrease	% Loss
		1,000 1-3/5 bushel boxes		
Brevard	302	56	246	–81.4
Charlotte	1,339	1,193	146	–10.9
Collier	529	503	26	– 4.9
DeSoto	150	69	81	– 5.4
Hardee	247	66	181	–73.2
Hendry	2,601	2,321	280	–10.7
Highlands	1,439	615	824	–57.3
Hillsborough	179	44	135	–75.4
Indian River	11,292	1,975	9,317	–82.5
Lake	437	412	25	– 5.7
Lee	316	289	27	– 8.5
Manatee	269	78	191	–71.0
Martin	1,058	274	784	–74.1
Okeechobee	470	379	91	–19.4
Orange	65	49	16	–24.6
Osceola	881	350	531	–60.0
Palm Beach	416	67	349	–83.9
Polk	3,121	1,207	1,914	–61.3
St. Lucie	15,359	2,613	12,746	–83.0
Sarasota	153	43	110	–71.9
Seminole	15	9	6	–40.0
Volusia	81	5	76	–93.8

Source: Florida Agricultural Statistics Service, Citrus Summary 2003-04 and Citrus Summary 2004-05.

TABLE 5-7. DECREASE IN TANGERINE PRODUCTION (ALL VARIETIES) BETWEEN THE 2003-04 AND 2004-05 SEASONS BY COUNTY.

County	2003-04	2004-05	Decrease	% Loss
		1,000 1-3/5 bushel boxes		
Brevard	39	15	24	–62
Charlotte	266	258	8	– 3
Collier	224	219	5	– 2
DeSoto	273	141	132	–48
Glades	84	72	12	–14
Hardee	276	138	140	–51
Hendry	499	539	+40[1]	+ 8
Hernando	12	7	5	–42
Highlands	565	392	173	–31
Hillsborough	176	96	80	–45
Indian River	162	58	104	–64
Lake	813	527	286	–35
Lee	93	73	20	–22
Manatee	63	30	33	–52
Marion	51	32	19	–37
Martin	98	77	21	–21
Okeechobee	104	82	22	–21
Orange	132	86	46	–35
Osceola	83	61	22	–27
Palm Beach	382	267	115	–30
Pasco	120	75	45	–38
Polk	1,460	1,011	449	–31
St. Lucie	463	154	309	–68
Sarasota	13	8	5	–38
Seminole	27	20	7	–26
Volusia	10	5	5	–50

[1]Hendry County showed a gain in tangerine production.
Source: Florida Agricultural Statistics Service, Citrus Summary 2003-04 and Citrus Summary 2004-05.

TABLE 5-8. CHANGE IN SPECIALTY FRUIT[1] PRODUCTION BETWEEN THE 2003-04 AND 2004-05 SEASONS IN MAJOR PRODUCING COUNTIES.

County	2003-04	2004-05	Change	% Gain or (Loss)
		1,000 1-3/5 bushel boxes		
Hardee	99	86	-13	(13)
Hendry	194	160	-34	(18)
Highlands	362	255	-107	(30)
Lake	197	373	+176	89
Polk	447	561	+114	26
St. Lucie	278	47	-231	(83)

[1]Includes Temples and Tangelos only.
Source: Florida Agricultural Statistics Service, Citrus Summary 2003-04 and Citrus Summary 2004-05.

which provided an estimate of the average fruit per tree; and the drop and size surveys, which provided estimates of the amount of fruit on the tree available at harvest and the size of the fruit at that time. The size of the samples used in these surveys and the distribution of the sample groves around the state were chosen to minimize error in the estimates of production and are not to be considered as accurate for the counties as at the state or area level.

Details by individual counties in citrus growing districts can be obtained from USDA, NASS, Florida Field Office reports or via the internet at www.nass.usda.gov/fl.

Further data by Florida commercial citrus production areas for the five seasons, 2000-01 through 2004-05 are shown in Tables 5-9 through 5-13 expressed in 1,000 1-3/5 bushel boxes. The counties which make up the five production areas are as follows:

Northern Area—Pasco, Hernando, Citrus, Sumter, Lake, Marion, Orange, Seminole, Sumter, western Volusia, and Flagler Counties.

Central Area—Polk, Osceola, and Highlands Counties.

TABLE 5-9. PRODUCTION OF ROUND ORANGES, GRAPEFRUIT AND SPECIALTY FRUIT, NORTHERN PRODUCTION AREA[1] 2000-01 THROUGH 2004-05.

	Northern Production Area		
Season	All Oranges 1,000 boxes	All Grapefruit 1,000 boxes	All Specialty[2] 1,000 boxes
2000-01	11,986	500	1,031
2001-02	13,526	478	1,495
2002-03	12,136	666	900
2003-04	13,551	595	1,477
2004-05	8,173	553	1,291

[1]Includes Alachua, Citrus, Lake, Marion, Hernando, Orange, Pasco, Putnam, Seminole, Sumter and part Volusia Counties.
[2]Includes tangerines, Temples, tangelos and K-Early citrus.
Source: Florida Agricultural Statistics Service, Citrus Summary 2000-01 through 2004-05.

TABLE 5-10. PRODUCTION OF ROUND ORANGES, GRAPEFRUIT AND SPECIALTY FRUIT IN CENTRAL PRODUCTION AREA[1] 2000-01 THROUGH 2004-05.

	Central Production Area		
Season	All Oranges 1,000 boxes	All Grapefruit 1,000 boxes	All Specialty[2] 1,000 boxes
2000-01	59,414	5,100	3,181
2001-02	66,074	6,222	3,993
2002-03	61,664	5,634	3,835
2003-04	75,149	5,405	2,971
2004-05	44,927	2,147	2,308

[1]Includes Highlands, Osceola and Polk Counties.
[2]Includes tangerines, Temples, tangelos and K-Early citrus.
Source: Florida Agricultural Statistics Service, Citrus Summary 2000-01 through 2004-05.

TABLE 5-11. PRODUCTION OF ROUND ORANGES, GRAPEFRUIT AND SPECIALTY FRUIT INDIAN RIVER AREA[1] 2000-01 THROUGH 2004-05.

Season	Indian River Area		
	All Oranges 1,000 boxes	All Grapefruit 1,000 boxes	All Specialty[2] 1,000 boxes
2000-01	19,100	30,700	1,439
2001-02	20,100	31,400	1,254
2002-03	16,000	24,100	1,230
2003-04	18,400	28,000	1,144
2004-05	6,200	4,800	307

[1]Includes Brevard, Broward, Indian River, part Martin, St. Lucie, part Palm Beach, and part Volusia counties.
[2]Includes tangerines, Temples, tangelos and K-Early citrus.
Source: Florida Agricultural Statistics Service, Citrus Summary 2000-01 through 2004-05.

Indian River Area—Eastern Volusia, Brevard, eastern Indian River, St. Lucie, eastern Martin and northeast Palm Beach Counties.

Western Area—Hillsborough, Manatee, Sarasota, Hardee and DeSoto Counties.

Southern Area—Charlotte, Lee, Glades, Okeechobee, Hendry, Collier, west Martin and west Palm Beach Counties.

Finally, the effect of the three, 2004 hurricanes on production statewide and on the most affected counties is summarized in Tables 5-14 thru 5-22.

TABLE 5-12. PRODUCTION OF ROUND ORANGES, GRAPEFRUIT AND SPECIALTY FRUIT IN WESTERN PRODUCTION AREA[1] 2000-01 THROUGH 2004-05.

Season	Western Production Area		
	All Oranges 1,000 boxes	All Grapefruit 1,000 boxes	All Specialty[2] 1,000 boxes
2000-01	60,200	1,500	1,022
2001-02	65,700	1,300	1,232
2002-03	54,000	1.700	1,466
2003-04	66,800	1,000	1,090
2004-05	31,800	300	696

[1]Includes DeSoto, Hardee, Hillsborough, Manatee, Pinellas and Sarasota Counties.
[2]Includes tangerines, Temples, tangelos and K-Early citrus.
Source: Florida Agricultural Statistics Service, Citrus Summary 2000-01 through 2004-05.

TABLE 5-13. PRODUCTION OF ROUND ORANGES, GRAPEFRUIT AND SPECIALTY FRUIT IN SOUTHERN PRODUCTION AREA[1] 2000-01 THROUGH 2004-05.

Season	Southern Production Area		
	All Oranges 1,000 boxes	All Grapefruit 1,000 boxes	All Specialty[2] 1,000 boxes
2000-01	72,600	8,200	2,317
2001-02	64,600	7,300	2,356
2002-03	59,200	6,600	1,719
2003-04	68,100	5,900	2,218
2004-05	28,800	4,100	2,048

[1]Includes Charlotte, Collier, Glades, Lee, Hendry, Okeechobee, part Martin and part Palm Beach Counties.
[2]Includes tangerines, Temples, tangelos and K-Early citrus.
Source: Florida Agricultural Statistics Service, Citrus Summary 2000-01 through 2004-05.

TABLE 5-14. PRODUCTION IN 1,000, 1-3/5 BUSHEL BOXES BY FRUIT TYPE AND CROP YEAR FOR 2000-01 THROUGH 2004-05.

	2000-01	2001-02	2002-03	2003-04	2004-05
Early-Midseason & Naval Oranges	128,000	128,000	112,000	126,000	79,100
Valencia Oranges	95,300	102,000	91,000	116,000	70,500
White Seedless Grapefruit	18,700	18,900	16,200	15,900	3,400
Colored Seedless Grapefruit	27,300[1]	27,800	22,500	25,000	9,400
Temples	1,250	1,550	1,300	1,400	650
Tangelos	2,100	2,150	2,350	1,000	1,550
Early Tangerines	3,550	4,350[2]	3,000	3,600	2,450
Honey Tangerines	2,050	2,250	2,500	2,900	2,000

[1]Excludes 2.0 million boxes of economic abandonment.
[2]Combined Dancy and Robinson varieties included in 2001-02 and discontinued in 2002-03.
Source: Florida Agricultural Statistics Service, Citrus Summary 2000-01 through 2004-05.

TABLE 5-15. EFFECT OF HURRICANES CHARLEY, FRANCES AND JEANNE ON FLORIDA CITRUS PRODUCTION, ALL VARIETIES.

Crop Year	Million boxes
2000-01	278.8
2001-02	287.3
2002-03	251.1
2003-04	291.8
2004-05	169.1

Source: Florida Agricultural Statistics Service, Citrus Summary, 2000-01 through 2004-05.

TABLE 5-16. EFFECT OF HURRICANES CHARLEY, FRANCES AND JEANNE ON FLORIDA ORANGE PRODUCTION.

Crop Year	Million boxes
2000-01	223.3
2001-02	230.0
2002-03	203.0
2003-04	242.0
2004-05	149.6

Source: Florida Agricultural Statistics Service, Citrus Summary, 2000-01 through 2004-05.

TABLE 5-17. EFFECT OF HURRICANES CHARLEY, FRANCES AND JEANNE ON FLORIDA GRAPEFRUIT PRODUCTION.

Crop Year	Million boxes
2000-01	46.0
2001-02	46.7
2002-03	38.7
2003-04	40.9
2004-05	12.8

Source: Florida Agricultural Statistics Service, Citrus Summary, 2000-01 through 2004-05.

TABLE 5-18. EFFECT OF HURRICANES FRANCES AND JEANNE ON ORANGE AND GRAPEFRUIT PRODUCTION IN ST. LUCIE COUNTY.

Crop Year	Oranges	Grapefruit
	- - - - - - Million boxes - - - - - -	
2000-01	9.97	17.03
2001-02	9.98	17.56
2002-03	8.26	13.28
2003-04	9.74	15.36
2004-05	3.50	2.61

Source: Florida Agricultural Statistics Service, Citrus Summary, 2000-01 through 2004-05.

TABLE 5-19. EFFECT OF HURRICANES FRANCES AND JEANNE ON ORANGE AND GRAPEFRUIT PRODUCTION IN INDIAN RIVER COUNTY.

Crop Year	Oranges	Grapefruit
	- - - - - - Million boxes - - - - - -	
2000-01	5.88	11.81
2001-02	6.67	12.14
2002-03	5.05	9.45
2003-04	5.72	11.29
2004-05	2.13	1.98

Source: Florida Agricultural Statistics Service, Citrus Summary, 2000-01 through 2004-05.

TABLE 5-20. EFFECT OF ALL THREE HURRICANES ON ORANGE AND GRAPEFRUIT PRODUCTION IN POLK COUNTY.

Crop Year	Oranges	Grapefruit
	- - - - - - Million boxes - - - - - -	
2000-01	29.31	3.12
2001-02	32.76	3.69
2002-03	30.68	3.41
2003-04	37.17	3.12
2004-05	21.00	1.21

Source: Florida Agricultural Statistics Service, Citrus Summary, 2000-01 through 2004-05.

TABLE 5-21. EFFECT OF HURRICANES CHARLEY, FRANCES AND JEANNE ON FLORIDA ORANGE AND GRAPEFRUIT PRODUCTION.

Crop Year	Oranges	Grapefruit
	- - - - - - - - - - - - Million boxes	- - - - - - - - - - - -
2000-01	223.3	46.0
2001-02	230.0	46.7
2002-03	203.0	38.7
2003-04	242.0	40.9
2004-05	149.6	12.8

Source: Florida Agricultural Statistics Service, Citrus Summary, 2000-01 through 2004-05.

TABLE 5-22. EFFECT OF HURRICANE CHARLEY ON COMBINED CITRUS PRODUCTION (ALL VARIETIES) IN CHARLOTTE, LEE, DESOTO AND HARDEE COUNTIES.*

Crop Year	Million boxes
2000-01	55.8
2001-02	59.2
2002-03	50.2
2003-04	61.8
2004-05	32.1

*While some damage may have occurred from Hurricanes Frances and Jeanne, it is estimated that most damage in these counties resulted from Charley.

Source: Florida Agricultural Statistics Service, Citrus Summary 2000-01 through 2004-05.

Chapter 6

THE EFFECT OF THE 2004 HURRICANES ON THE ORANGE JUICE FUTURES MARKET

Trading in orange juice futures began in 1966. The unit of trading was 15,000 pounds of orange solids (3% more or less). It became immediately apparent to processors, growers and speculators that the market would be subject to the vagaries of severe weather, particularly freezes and hurricanes. Strangely enough, until 2004 no hurricane significantly affected the Orange Juice Futures Market. Occasionally, the presence of a hurricane in the Atlantic, the Caribbean or the Gulf of Mexico would lead to an upturn in the market, but in all cases the perceived threat fizzled and speculators began to regard hurricanes as opportunities for short sales.

Historically, Florida had experienced hurricanes in October 1944, August 1949 and Donna in September 1960, all three of which would have had a serious impact on the Orange Juice Futures Market if it had existed (Attaway, 1999). Losses of all varieties of citrus in the 1944 hurricane were estimated at 23 million 1-3/5 bushel boxes from a 69 million box crop. The 1949 hurricane reduced a projected 88 million box crop by 15 million boxes and Donna in 1960 destroyed 22-24 million boxes of all varieties from an expected 124 million box crop (Attaway, 1999). However, it should be noted that the larger part of these losses on a percentage basis was to grapefruit rather than oranges (Table 6-1).

One should also note that major hurricane damage was usually limited to the coastal counties as peak winds decreased as the storms moved inland. For example, the 1944 and 1960 hurricanes were most destructive to groves in the southwest Florida counties such as Lee, Charlotte and DeSoto, but also did major damage in Hardee, Highlands and Polk counties while the 1949 hurricane influenced the Indian River area much more than the interior counties. Greatest damage was in Martin, St. Lucie and Indian River counties, but significant damage also resulted in Polk County.

Other hurricanes, from 1928 to 1979, which had a small but a measurable affect on the Florida citrus production, are shown in Table 6-2.

On the other hand, damage from freezes was much more widespread than hurricane damage. For example, freezes in January 1977, January 1981, January 1982, December 1983, January 1985 and December 1989, produced wide upswings in the Orange Juice Futures Market (Attaway, 1997). Unfortunate speculators who were short in the futures market suffered severe losses.

TABLE 6-1. ESTIMATED LOSSES OF GRAPEFRUIT AND ORANGES DURING THE HURRICANES OF 1944, 1949 AND 1960.

Hurricane Year	Oranges	Grapefruit
	Million boxes	
1944	13.7	9.2
1949	11.5	3.6
1960	14.0	7.0

Source: Citrus Industry, 1960.

Those fortunate enough to be on the long-side of the market made substantial profits.

Examples of market shifts after severe freezes are shown in Table 6-3.

As it resulted in massive imports of orange juice from Brazil, the freezes of the 1980s changed the Orange Juice Futures Market forever. It was felt by many that without imports from Brazil the 1983 and 1985 freezes would have pushed OJ prices over $2.00/pounds solids.

The Futures Market Prior to Hurricane Charley

In the months preceding Hurricane Charley, the U.S.D.A. Crop and Livestock Reporting Service (Florida Field Office, National Agricultural Statistics Service. June 10, 2004) estimate for the 2003-04 crop was holding at 245 million boxes, up from 203 million boxes a year earlier. Orange juice prices were approaching record lows. On May 21, 2004, a contract for July delivery on the New York Cotton Exchange dipped to 54.2 cents per pound solids, the lowest since 1977. The daily volume was less than 2,000 contracts with open interest at 35,982 contracts.

Dr. Dan L. Gunter, Executive Director for the Florida Department of Citrus, attributed the collapse in prices to the largest crop in history, and a massive carryover of juice from the previous year combined with a 5% fall in consumer demand.

On June 10 the futures price for July delivery again dropped below 55 cents, with 13% of the 2003-04 Valencia orange crop still on the trees, up from 4.8% in June 2003. However, the market began a gradual increase as the daily volume rose from 4,551 contracts to 11,773 contracts on June 24 with the futures price increasing to

TABLE 6-2. OTHER HURRICANES WHICH AFFECTED ALL VARIETIES OF FLORIDA CITRUS PRODUCTION AND THE COUNTIES MOST AFFECTED.

Date	Total Crop	Estimated Loss	Counties
	------- Million boxes -------		
Sept. 16-18, 1928	27.8	1.0	DeSoto, Hardee & Polk
Sept. 4-5, 1933 (Labor Day Hurricane)	28.8	4.5	Indian River, St. Lucie, Martin, Palm Beach and Highlands
Sept. 15-16, 1945	86.2	4.0	Highlands
October. 6-8, 1946	83.3	2.0	Lee, Charlotte & Sarasota
Sept. 17-18, 1947	91.3	6.0	Lee & DeSoto
Oct. 4-5, 1948	93.1	2.0	Lee, Charlotte & Sarasota
Oct. 17-18, 1950 (Hurricane King)	105.4	2.5	St. Lucie & Indian River
Aug. 27-28, 1964 (Hurricane Cleo)	123.7	3.0	Martin & St. Lucie
Sept. 3-4, 1979 (Hurricane David)	283.6	5.0	Martin, St. Lucie & Indian River

Source: Climatological Data, Florida Section, 1928-1979 and Florida Agricultural Statistics Service, 1997.

60 cents per pound of solids, July delivery. Open interest, defined as the total number of futures contracts or option contracts that have not been exercised, expired or fulfilled by delivery, was 43,823 contracts.

When the U.S.D.A. (Florida Field Office, National Agricultural Statistics Service) made its final estimate of the 2003-04 crop on July 13, the size of the orange crop was reduced from 245 to 242 million boxes, down 1% from June estimates. In the interim on July 12, the contract price had risen to 65.20 cents with a volume of 2,493 contracts on open interest of 40,185 contracts.

When trading ended in the July contract, September became the near month and closed at 68.10 cents per pound of solids on July 13. During the balance of July, the September contract traded as high as 72 cents on the 26th of the month with a volume of 12,252 contracts to a low of 60.60 cents on the 30th of the month with only 1,356 contracts traded.

Hurricane Charley

Meanwhile, on August 4 a tropical wave moved off the coast of Africa and began to drift slowly westward toward the Caribbean Sea where it became Tropical Storm Charley on August 10 and minimal hurricane Charley on the 11th. Strengthening south of Cuba, Charley reached Category 2 status on the Saffir-Simpson Scale on Thursday August 12. The futures market took no notice, trading from 58.75 to 60.90 cents and closing at 60.00 cents with a volume of 6,019 contracts.

When a serious weather event is approaching, Friday is always a critical day for the Orange Juice Futures Market. Friday, August 13, 2004, was an exceptionally critical day but the market **took no notice**. Trading ranged from 63.50 to 59.30 cents with a volume of only 5,499 contracts, 520 contacts less than the previous day. Why? Two reasons are obvious. One, the National Hurricane Center forecast indicated that Charley would parallel the coast toward the Tampa area and on to north Florida, sparing the bountiful groves of southwest and central Florida. Two, a hurricane had never seriously affected the Orange Juice Futures Market since trading began in 1966. A third factor may have been an estimate by Louis Dreyfus Citrus for a 230 million box crop in 2004-05 and an unofficial estimate by the Florida Citrus Commission for 226 million boxes.

As the market was up two cents on the 13th, some speculators may have taken advantage to sell short on Friday afternoon. If so, this was a serious mistake because after the market closed for the week, Charley turned to the northeast over Punta Gorda, Arcadia and points north. The rest is history. As damage reports poured in over the

TABLE 6-3. EFFECT OF MAJOR FREEZES ON THE ORANGE JUICE FUTURE'S MARKET.

Freeze Year	Price of OJ Prior to Freeze	Peak price of OJ After the Freeze
1977	38.5¢/pounds solids	79.75¢/pounds solids
1981	78.4¢/pounds solids	143.75.¢/pounds solids
1982	118.35¢/pounds solids	152.4¢/pounds solids
1983	122.55¢/pounds solids	170.00¢/pounds solids
1985	164.5¢/pounds solids	183.¢/pounds solids
1989	121.4¢/pounds solids	205.3¢/pounds solids

Source: J. A. Attaway, 1997. Citrus Associates, New York Cotton Exchange.

weekend, traders could do nothing but wait for the market opening on Monday morning, August 16th, and on Monday there was no relief with 9,380 contracts trading as high as 72.25 cents, a "dime" higher than Friday's close. A bull market had begun, with daily trading volume exceeding 10,000 contracts on August 19th and 20th.

Hurricane Frances

For the balance of August the market traded mostly between 72 and 75 cents per pound of solids with decreasing volume to as low as 2,399 contracts on Friday August 27. Then, trouble again brewed as a new Cape Verde hurricane formed in the Atlantic and reached 140 mph, category 4 status on the Saffir-Simpson Scale. On August 31, 2004, it passed north of the Virgin Islands and moved into the Bahama Islands on the second of September. This hurricane was named Frances, and although she weakened to a Category 2 hurricane, with 90-100 mph winds at landfall near Stuart, she moved westward through Martin and St. Lucie counties causing major damage to the heart of the Indian River District. Significant damage also occurred in Polk and Highlands counties. The futures market no longer had a blase attitude toward a hurricane, and on September 2, 2004, the price peaked at 85 cents per pound of solids. The market would see only two trades below 70 cents for the balance of 2004 as the bull market in orange juice persisted into 2005.

Hurricane Ivan

Hurricane Ivan followed closely after Hurricane Frances, but did not strike the Florida citrus belt. Ivan was a Cape Verde hurricane which actually attained category 5 status on the Saffir-Simpson Scale as it passed into the Caribbean Sea on September 8. Ivan passed south of Jamaica as a category 4 hurricane then re-strengthened to category 5 status on September 11 and moved west-northwest to devastate the Cayman Islands. On September 13 Ivan passed into the Yucatan Channel and emerged into the Gulf of Mexico on September 14. Moving north the threat of Ivan to the citrus belt diminished as it continued toward the Alabama-Florida panhandle border as a category 3 hurricane. On September 9th and 10th the futures market traded between 83 and 81 cents per pound of solids. Ivan may have had some effect, but the direct hit by Hurricane Frances on the groves in Martin, St. Lucie and Indian River counties doubtless had the major influence.

Hurricane Jeanne

The third blow to the citrus belt was Hurricane Jeanne which closely followed the track of Hurricane Frances, making landfall just east of Stuart, Florida on September 26 with 120 mph winds, category 3 on the Saffir-Simpson Scale. Jeanne moved inland toward central Florida becoming the third hurricane to pass through Polk County in the same year. Jeanne's winds were stronger than Frances and she made a major contribution to the loss of fruit and trees, and had a significant effect on the futures market. From September 27 to October 20, the Orange Juice Futures Market, now for November delivery, did not close below 82 cents per pound of solids, trading in a range from 82 to 89 cents.

In Summary

In the spring and early summer of 2004, orange juice futures were in a bear market, trading from 54.2 cents per pound of solids, the lowest in 27 years, to a mid 60 cent range. After three hurricanes, the high and low was in the mid 80 cent range, 20 cents higher. After dropping into the 70 cent range in mid-November and early December, the market again returned to the 82-88 cent range which held until the end of December. The bull market continued at this level until March 10, 2005, when it soared over one dollar per pound of solids. After that it traded from the mid 90 cent to the dollar level before closing at 96.35 cents on June 30, 2005.

As a result of the damage which resulted from the three hurricanes, the value of an orange juice futures contract had almost doubled from 55 cents per pound of solids in June 2004 to 101.70 in June 2005.

Futures traders will not ignore hurricanes in the future.

Terms Related to the Orange Juice Futures Market

Bear Market—A prolonged period during which juice prices fall accompanied by widespread pessimism.

°Brix—The unit of measure for soluble solids.

Bull Market—A prolonged period during which juice prices rise accompanied by widespread optimism.

Open Interest—The total number of juice future's contracts in a delivery month that has been entered into but not yet liquidated by an offsetting transaction or fulfilled by delivery.

Orange Juice Future's Contract—A contract consists of 15,000 pounds of solids. The price of a contract can be calculated by multiplying the price per pound solids by 15,000.

Orange Juice Future's Market—Orange juice futures were previously traded on the New York Cotton Exchange. Effective June 10, 2004, Orange Juice Futures are traded on the New York Board of Trade.

CHAPTER 7

EVENTS AFTER THE 2004 HURRICANES

The Federal Disaster Program for 2004 Hurricanes

Immediately after Hurricanes Charley and Frances had stormed across Florida leaving behind massive damage to the state's agriculture, President George W. Bush declared counties in the path of the two hurricanes to be in a disaster area. In response to a directive from the President to the USDA, Secretary of Agriculture Veneman announced in Bartow, Florida on September 24, 2004, that producers of citrus, fruits and vegetables and nursery crops located in Florida counties in the disaster area would be eligible for special disaster relief. The announcement was made in cooperation with Governor Jeb Bush and Representative Adam Putman. Secretary Veneman further added that "President Bush is committed to relief for Floridians and existing funds would be used to quickly launch the Florida Hurricane Agricultural Disaster Assistance Program in an effort to help growers in their cleanup efforts and compensate them for lost crops and trees." She added further "that the USDA would continue to work with Congress to ensure adequate funding for hurricane relief."

In a release on the same date from Governor Jeb Bush's office the Governor thanked the USDA for providing $500 million in hurricane disaster assistance for producers of citrus, fruits and vegetables and nursery crops. The Governor added that "Our federal partners have been extremely responsive to the needs in our state on a variety of fronts, and I thank the USDA for its rapid response to put funds in the hands of farmers so they can get back on their feet," adding that, "the relief provided today will go a long way in providing the critical emergency support, and I appreciate the willingness of USDA and Congress to ensure adequate funding for hurricane relief."

The Governor announced that the state will provide $10 million in funds to help cover the shortfall in box taxes which go to the Florida Department of Citrus (FDOC) to promote Florida citrus and remind the market that Florida's agriculture industry still offers the best produce and products in the country.

Immediately following the September 24 announcements from the offices of Secretary Veneman and Governor Bush, the citrus industry was struck on September 26 by an unprecedented third hurricane, Hurricane Jeanne, and the disaster assistance program was broadened to include damages resulting from Hurricane Jeanne. In a fact sheet from the USDA Farm Service Agency it was noted that payments for this assistance are authorized under Section 32 of the Agricultural Act of August 24, 1935, which allows the Secretary of Agriculture to restore producers' purchasing power. Aid to producers under this program is estimated at more than $500 million.

Eligible counties were: Alachua, Baker, Bradford, Brevard, Broward, Charlotte, Citrus, Clay, Collier, Columbia, DeSoto, Dixie, Duval, Flagler, Gadsden, Gilchrist, Glades, Hamilton, Hardee, Hendry, Hernando, Highlands, Hillsborough, Indian River, Jefferson, Lafayette, Lake, Lee, Leon, Levy, Liberty, Madison, Manatee, Marion, Martin, Miami-Dade, Monroe, Nassau, Okeechobee, Orange, Osceola, Palm Beach, Pasco, Pinellas, Polk, Putnam, Sarasota, Seminole, St. Johns, St. Lucie, Sumter, Suwannee, Taylor, Union, Volusia and Wakulla.

Citrus sign-up was to begin on October 5, 2004 with citrus producers to be reimbursed on a per acre basis for each eligible grove. Payments were based on the severity of destruction as determined by the path of the storms and damage estimates developed by the Farm Service Agency (FSA) in cooperation with the Florida Department of Agriculture. The formula took into account various levels of losses generally correlating to the distance from the eye of the hurricanes, the average production loss, tree loss, and rehabilitation and cleanup costs. Fruit from the groves must have been marketed in both 2003 and 2004. Payment rates per acre for participants with insurance or Noninsured Crop Disaster Assistance Program (NAP) coverage was as follows:

Tier 1
75% or greater crop loss and associated tree damage = $1,500

Tier 2
40-74% crop loss and associated tree damage = $1,000

Tier 3
35-49% crop loss and associated tree damage = $600

Tier 4
>15% tree damage = $100

Each producer's payment was limited to one $80,000, subject to limitation under the program payment rates, but would be 5 percent less for producers who did not obtain Federal Crop Insurance. Insurance was available from the Risk Management Agency, or on coverage under the NAP, available from the Farm Service Agency. Similar to a number of other USDA programs, payments would not be available for producers whose adjusted gross income is $2.5 million or higher, unless 75 percent or more was derived from farming and forestry. In addition, producers would be required to agree to purchase crop insurance or NAP coverage for next year's crop.

All producers applying for benefits must certify that they did indeed suffer the level of loss claimed and that the losses were caused by Hurricanes Charley, Frances or Jeanne.

President Bush Visits the Florida Citrus Industry

Three days after Hurricane Jeanne had made landfall, and at an invitation from Florida Citrus Mutual, President Bush made a personal visit to groves near Lake Wales owned by Mutual President Marty McKenna and his brother Pat McKenna. Located in Polk County, the McKenna groves received direct hits from three of the 2004 hurricanes, Charley, Frances and Jeanne. The President's 30 minute tour of the groves occurred on September 29, 2004 and was described by Kevin Bouffard in his Ledger byline entitled *Growers Recall Bush's Charm, Citrus Knowledge.* "I was tickled to death he called me 'Marty'," McKenna said and the day after the visit brother Pat McKenna recalled "I'm still pinching myself that it really was the President kicking dirt with us." The President was accompanied by his brother, Florida Governor Jeb Bush, Secretary of Agriculture Ann Veneman and Florida Commissioner of Agriculture Charles Bronson.

The President's first question, after seeing all the oranges on the ground, concerned how much damage each of the hurricanes caused. Pat McKenna replied "that the groves were 'color-coded': the oranges which were brown and black fell during hurricane Charley on August 13. The lime green oranges fell during Frances over the Labor Day weekend, and the shiny, dark green fruit had dropped during Jeanne over the previous weekend."

The McKennas told Bush they estimated they had lost 40 percent of the citrus crop in the groves they managed in Polk, Highlands and Hardee counties, from Hurricanes Charley and Frances, with additional droppage expected from Hurricane Jeanne. The brothers said that they were impressed with the President's knowledge of agriculture in general. Marty McKenna indicated that the President was knowledgeable about the land, the environmental conditions and how the losses would affect the people.

The McKennas are partners in McKenna Brothers, Inc., a grove service company that manages more than 5,000 acres in central Florida. Florida Citrus Mutual had given the White House three possible sites for the President to inspect grove damage from hurricanes Charley, Frances and Jeanne. The McKennas were not informed that their grove off State Road 60, about 5 miles east of Lake Wales was to be the tour site until less than 24 hours before the President's arrival.

Labor Day, September 5, 2005— The Epilogue

In Chapter 2, (Hurricane Frances) and Chapter 4, (Hurricane Jeanne), Peter Spyke of Arapaho Citrus Management in Fort Pierce provided e-mails describing his personal experiences and the experiences of his family during the two storms. The following epilogue describes his observations a year later.

It's the one year anniversary of Hurricane Frances. The newspaper ran a special section this morning with some photos of places last year and this year to show the recovery. As I looked at the pictures, I noticed that tears came to my eyes as I thought back to that day. I guess that means that the emotional trauma is not completely gone, which is understandable.

It has been an interesting year, filled with high and low points. Unfortunately, at the end of the day, it looks like the storms had an impact that none of us could ever have predicted. We're still not completely recovered. We managed to get a new roof on the grove house when Darwin, a friend of Cindy's best girlfriend, came down from New York state to fix her roof. We met Darwin down at Cindy's friends house and talked him in to coming up to Ft. Pierce and fixing the roof on the office. He stayed for about 3 weeks, and put a new roof on the house and also rebuilt the pole barn. We used architectural shingles on the house, and Darwin built the barn back stronger than before. It cost us the deductible on the insurance, and I suppose we are better off now. In the process, we made a new friend in Darwin, so that's one of the blessings embedded in the whole thing.

One of the rooms in the Ft. Pierce house had the ceiling collapse because it had gotten wet, so there's a big hole in the ceiling and so far I haven't found someone to fix that. Darwin patched a bunch of rotten wood when he replaced the roof, plus rebuilt the carport that had blown off, so that has to be sealed and painted. The entire outside of the Ft. Pierce house needs to be patched and painted. We drilled so many holes in the stucco to put up plywood that it's a real job. I'm waiting until summer is over so everything can dry out before we do that work.

The Orange Shop's season was so-so last year, which I guess is a blessing, too—we had a year to be so-so. It turns out that there was so much publicity about the citrus losses in the hurricanes that a lot of customers ordered something other than fruit for their holiday gifts. Plus, traffic was down early in the year because a lot of the hotels were full of hurricane refugees. When they did finally make it down, they were surprised that we had fruit. Plus, the fruit was really expensive, so despite price increases our margins were thin. Things finally picked up later in the season, so we were able to salvage things fairly well.

Between fixing up the groves and the marginal year at The Orange Shop, the crop and disaster money we received from the government is used up. Thank goodness we had it, though, or we would have been dead in the water. I did go ahead and push up Rock Bottom and part of my 36 grove since the trees were so damaged from the flooding that they weren't going to come back. The rest of the groves came back with about a 60-70% crop, so we were looking forward to being able to have a decent year because fruit prices are very high.

Until Citrus Canker, that is. As it turns out, the storms did wipe us out—it just took a while to learn that it had happened. A number of years ago, Citrus Canker had been discovered in Miami. Canker is a very serious disease, and whenever it's found we try to eradicate it rather than just control it. Since it was first discovered, it had spread through the dooryard citrus trees in Dade and Broward Counties, but had only popped up here and there in commercial groves.

When it is found in commercial groves, the protocol is to remove the infected trees plus all the trees within another 1,900 feet since they are considered to be exposed. Each infected tree therefore generates about a 240 acre circle that must be removed.

The disease spreads naturally by wind-blown rain. It's a bacterium that colonizes the leaves and fruit, causing a blemish. It's almost impossible to control once a tree has been infected.

There had apparently been a grove in southern Indian River County that had canker at the time of the storms. Between Frances, Jeanne, and Ivan, canker from that grove spread for 40 miles. Some was found in Shadow Lakes, and three of my neighbors in Ft. Pierce have groves in which it was found that are within 1,900 feet of my groves. So, as I write this, all of Shadow Lakes will have to be removed, and all of my trees except for 10 acres of Lemons will be removed. There will be compensation, but it may not come for 2-3 years.

The area of infection in St. Lucie and Martin Counties has at this point required that about 18,000 acres of citrus be removed. There are a lot of people who won't replant. I will probably try to replant the Home Grove and Rock Bottom at least, provided that I can survive the financial challenges. Shadow Lakes was our primary caretaking customer, and a fair number of other customers have also lost their groves. The industry is reeling, and none of us really know what's going to happen since they are finding more canker every day as it's being spread by our summer rains.

Land prices are fairly high, so we may need to sell some land to make ends meet. Since I had already pushed out most of 36 and Rock Bottom before canker was found in the neighbor's groves, I won't receive compensation for those trees and we have to wait 2 years before we replant. After that, it's about another 5 years before the trees begin to produce fruit, so that's a long haul with no cash flow.

Thank goodness for The Orange Shop. Both my caretaking business and personal groves are floundering, but The Shop still should be OK. I can't imagine what it would be like without some kind of operating business, so we're focusing on the Shop and trying to build that a bit more than we would have done otherwise. The Shop's Navel grove has a good crop and no canker, and we bought the Navel grove in Weirsdale for the family after we sold a piece of land in Ft. Lauderdale. So, we're still in the fruit business, and there is something to build on.

I'm pretty sure we'll be OK financially, one way or the other, although some of the decisions may be painful. There will be jobs lost, too. After the storms reduced the crop last year, the canker is doing it again this year. St. Lucie County will lose about 25% of the citrus, which means 25% of the crop, plus the crops are light this year after the storm damage from last year. A lot of grove workers, pickers, and packinghouses will be hurting due to the lack of fruit, and there's no disaster money coming in this time.

We're determined to try and keep going. We're looking at tropical fruits and vegetables as alternative crops. The problem is that we don't know much about those industries—they're very different from citrus. We have land and good people, though, so that's a significant set of assets. Hopefully we will be able to find the positive choices, and follow a path that leads us to a healthy, happy place.

It has been one terrible year. Other than having health problems, I couldn't imagine things being much worse until Hurricane Katrina hit the Gulf coast this year. At least we had resources such as pumps, cars, roofs, and so forth. We lost a lot, but it didn't all happen at once so we had chances to regroup emotionally. At the end of the day, we're certainly a lot better off than a lot of those people, and even a lot of people around here who lost their homes and all their belongings.

I think the big difference was the flooding. Our water could be pumped off or drain off since most of the land is still in agriculture and we had facilities to remove water on an individual basis. If your neighbor's pump had a problem, you could still pump out your own water. That allowed us to get rid of the water that has caused so many problem in New Orleans.

Just goes to show you that no matter how bad things get, there's still something to give thanks for. We just have to focus on our blessings.

Post-Hurricanes Update—Selected Counties

On September 16, 2005, in response to the need for more accurate citrus tree numbers following the hurricanes of 2004, the National Agricultural Statistics Service (NASS) conducted a special census in selected counties during the first six months of the year. Two counties, DeSoto and Hardee, are in the Western growing area and were subjected to strong winds from Hurricane Charley. The other four counties—Indian River, Martin, Palm Beach and St. Lucie, are along the Atlantic coast, and bore the brunt of Hurricanes Frances and Jeanne.

Field personnel updated records by conducting field enumeration, visiting grove managers, and consulting with the Division of Plant Industry's Citrus Canker Eradication Program. It is important to note that tree losses are not due solely to hurricane damage, but also reflect effects of the citrus tristeza virus, urban development, and removals due to citrus canker through June 2005.

Acreage losses from 2004 to 2005 vary from 1,486 (Palm Beach) to 18,560 (St. Lucie) Table 7-1. Hardee's acreage peaked in 2002, however DeSoto's high was in 2000. Acreage in Indian River, Martin and St. Lucie have been trending downward since 1994, while Palm Beach has been declining since 1970. Comparisons between 2004 and 2005 for each fruit category show that decreases in acreage were larger than corresponding decreases in tree numbers, with acres of oranges down 10.7%, grapefruit 18.5%, specialty fruit 16.4% and all citrus 12.7%.

Palm Beach had the fewest trees in the selected counties and the number of trees has been reduced by one-third. Tree counts in DeSoto, Hardee and Indian River are down over 500,000 trees each. St. Lucie's total is down more than 2.1 million trees with additional losses to canker occurring after the survey period. Table 7-2.

The loss of trees in the citrus inventory by acre and number of trees, from 2004-2005, is shown by fruit variety in Tables 7-3 and 7-4.

TABLE 7-1. TOTAL CITRUS: NUMBER OF ACRES BY CENSUS YEAR AND COUNTY.

	2000	2002	2004	2005
County	Acres			
DeSoto	71,781	70,365	68,559	63,938
Hardee	53,115	54,961	54,414	48,843
Indian River	60,293	56,012	47,539	42,916
Martin	44,746	42,208	40,330	37,342
Palm Beach	10,090	7,964	4,542	3,056
St. Lucie	98,899	92,490	82,987	64,427
Total	338,924	324,000	298,371	260,522
Decrease from previous inventory		4.4%	7.9%	12.7%

Source: National Agricultural Statistics Service, Florida Field Office, Orlando, FL.

TABLE 7-2. TOTAL CITRUS: NUMBER OF TREES BY CENSUS YEAR AND COUNTY.

	2000	2002	2004	2005
County	1,000 trees			
DeSoto	9,369.6	9,282.3	9,080.1	8,541.6
Hardee	6,116.5	6,466.7	6,462.6	5,958.1
Indian River	6,583.6	6,191.3	5,322.3	4,821.7
Martin	6,191.2	5,947.6	5,732.2	5,365.4
Palm Beach	1,361.2	1,128.8	699.8	468.1
St. Lucie	11,837.8	11,266.2	10,342.0	8,191.0
Total	41,459.9	40,282.9	37,639.0	33,345.9
Decrease from previous inventory		2.8%	6.6%	11.4%

Source: National Agricultural Statistics Service, Florida Field Office, Orlando, FL.

TABLE 7-3. CITRUS: NUMBER OF ACRES BY COUNTY, TYPE AND CENSUS YEAR.

	Oranges		Grapefruit		Specialty		Total	
	2004	2005	2004	2005	2004	2005	2004	2005
County	Acres							
DeSoto	67,020	62,386	438	615	1,101	937	68,559	63,938
Hardee	52,384	47,095	620	494	1,410	1,254	54,414	48,843
Indian River	21,901	19,715	24,194	21,918	1,444	1,283	47,539	42,916
Martin	36,373	34,038	2,823	2,388	1,134	916	40,330	37,342
Palm Beach	2,176	1,084	779	584	1,587	1,388	4,542	3,056
St. Lucie	40,259	32,211	37,990	28,450	4,738	3,766	82,987	64,427
Total	220,113	196,529	66,844	54,449	11,414	9,544	298,371	260,522
Percent Decrease		10.7%		18.5%		16.4%		12.7%

Source: National Agricultural Statistics Service, Florida Field Office, Orlando, FL.

TABLE 7-4. CITRUS: NUMBER OF TREES BY COUNTY, TYPE AND CENSUS YEAR.

	Oranges		Grapefruit		Specialty		Total	
	2004	2005	2004	2005	2004	2005	2004	2005
County	Trees							
DeSoto	8,868.5	8,325.0	51.2	78.4	160.4	138.2	9,080.1	8,541.6
Hardee	6,209.9	5,734.4	67.2	57.4	185.5	166.3	6,462.6	5,958.1
Indian River	2,584.9	2,342.2	2,544.1	2,305.9	193.3	173.6	5,322.3	4,821.7
Martin	5,254.4	4,958.2	314.5	270.5	163.3	136.7	5,732.2	5,365.4
Palm Beach	352.4	174.3	83.8	64.2	263.6	229.6	699.8	468.1
St. Lucie	5,522.7	4,533.6	4,170.8	3,141.3	648.5	516.1	10,342.0	8,191.0
Total	28,792.8	26,067.7	7,231.6	5,917.7	1,614.6	1,360.5	37,639.0	33,345.9
Percent decrease		9.5%		18.2%		15.7%		11.4%

Source: National Agricultural Statistics Service, Florida Field Office, Orlando, FL.

CHAPTER 8

THE AUTHOR'S NEIGHBORHOOD

The Author's home in Polk County is on the east side of Lake Otis which in turn is on the east side of Winter Haven in a neighborhood called Lakewood. As this was the weaker side of Hurricane Charley, sustained winds were only near 50-60 mph, with peak gusts to 70-75 mph, less than the 105 mph recorded at the Florida's Natural Processing Plant in Lake Wales which took a direct hit.

On the evening of August 13 we were eating dinner in the sunroom with our daughter Marie and grandson Max who had left their home in Tampa as early predictions had indicated that Charley would strike the Tampa Bay area rather than Polk County. Many Tampa families were scattered up and down I-4 on the way to Orlando. At 7 p.m. the winds hit tropical storm force in Winter Haven and limbs began to crash from the giant oak tree in the back yard into the swimming pool. The oak tree was split and had to be removed after the hurricane.

Lakewood had laurel oaks approaching 40 years in age which were very vulnerable as several lined a narrow, two lane causeway between Lake Otis and Lake Link with no protection from the winds. The causeway was the only access to the neighborhood. Laurel oaks are normally on the downside of their life cycle after 40 years, and the Lakewood oaks were no exception. Over thirty, 40-year old oak trees were lost. Many had fallen across streets or roof tops. Figures 8: 1-8 show damage around the neighborhood.

Lakewood residents were fortunate to have a Community Association chaired by Bill Watson, a native of the Winter Haven area. Although the National Hurricane Center was forecasting landfall in the Tampa Bay area, Watson recognized that our area would be subject to major damage if the hurricane passed through Polk County. Two days before Charley struck he reserved a tree crew and crane. Reserving the crew was even more important as they arrived at sunup on August 14 and began the job of opening up the causeway which was totally blocked by trees uprooted by wind gusts blowing across Lake Link from the east. Without Watson's foresight, residents would have been unable to leave the area for a week or more, except on foot or on a bicycle, and emergency services (EMT) could not have entered the neighborhood. The causeway had also been hit by a small tornado spawned by the hurricane which aggravated the damage to the trees and the road.

Other neighborhood residents deserving of credit were the "first hour group" who attacked the downed trees with chainsaws as soon as the storm subsided. This group, who worked all night and until 11:00 a.m. the next morning, made it possible to move around the neighborhood while waiting for the causeway to be opened by the tree crew and crane. Making up the "first hour group", were Hood Craddock, Ron Brooker, Scott Lingenfelter, Mark McCollough, Steven Daniel, Dr. Bill Myers, C. Ober, Rodney Heintz, Larry Helms, Ron Kirin, Tom Lindsey and Bill Watson. The author apologizes for anyone who may have been left out.

Many interesting things were learned during the passage of three hurricanes, one being that it did not require the uprooting of a tree to break a water line. During the shaking of trees by the long hours of high winds, water lines were broken by the continuous moving of the root systems. This happened at several locations.

The following newsletter was sent to neighborhood residents by the Lakewood Community Association on September 8, 2004 after Hurricane Frances.

I hope this finds everyone in some state of recovery and celebration that we have made it through our second hurricane! If we can all be patient and understanding with our neighbors and friends during this time of high emotional stress, we will be okay.

In regards to being okay:

The water service on Lake Otis Road, Park Lane and Lake Mariam Drive should hopefully be restored this evening. The city is working hard to fix the most current line break. This now totals **fourteen line breaks** in this neighborhood alone. All have been a result of trees shifting and or falling due to the high winds of the two hurricanes. Please keep in mind that interruptions to your water service may continue until all cracks and breaks are detected and repaired.

The City of Winter Haven Water Department is working around the clock. Stop calling the City of Winter Haven Water Department UNLESS you notice a fresh break.

Due to the fact that additional water line breaks are likely, you may want to keep up with laundry and dishes in the event you find yourself without water, yet still again.

We, as a community, are trying to keep our roads as clear and as safe as possible. Please make all attempts to clear the drains by your home to prevent flooding and do not stack your yard debris on the road. If you happen to have a fire hydrant in your yard, please make sure it is clear as well. We are in con-

Figures 8: 1-8 show photographs taken in Lakewood over the days immediately following Hurricane Charley.

tact with the county with regard to picking up our debris, but we do not have a commitment from them on when they will be in our neighborhood. We do believe they will be here next week. If you want to contract with someone to remove your debris, please do so.

The Lakewood Community Association is making all attempts to expedite the recovery of our neighborhood to return it to its splendor. We thank all those homeowners who have paid their special assessment and encourage those who have not paid to pay now or contact us to discuss a payment plan.

The following Precautionary Boil Water Notice was issued by the City of Winter Haven on September 7, 2004, immediately after Hurricane Frances:

City of Winter Haven Utility/Engineering Services' Department is issuing a precautional boil water notice due to disruption in water service in your immediate area.

As a precaution, it is advised that all water used for drinking or cooking be brought to a full rolling boil for at least one minute. The city is also advising that water which had been previously boiled be used to wash fruits and vegetables, make ice and wash hands, etc.

This precautionary boil water notice will remain in effect until ongoing water sampling analyses are completed. As a precaution, the water lines in the area have been flushed. City of Winter Haven Utility/Engineering Services' Department will issue a notice to residents rescinding this precautionary measure once satisfactory results have been confirmed. Thank you in advance for your cooperation on this matter. For further questions, please contact the Fairfax Water Treatment plant or the Utility Maintenance/Development Division.

And on September 11, 2004, the city issued the following notice rescinding the Precautionary Boil Water Notice.

This is to inform you that the precautionary boil water notice recently hand-delivered to your residence/business is hereby rescinded. Satisfactory results were obtained from water samples recently collected.

The City of Winter Haven appreciated your patience during this period and regrets any inconvenience it might have caused.

A year and two months after Hurricane Jeanne, the last of the three hurricanes to hit our neighborhood, the effect of the hurricanes is very much in evidence. The leaves on the tall oaks on the high ground, which were defoliated, are very bare, particularly in the tops of the trees. Several trees are so bare that they will probably have to be removed. Trees which were pruned stand tall and thin but the new leaves are bright and healthy. These trees will probably live for many more years.

The sod on our quarter-mile causeway between Lake Link and Lake Otis had been washed out by Charley, but new sod 30 feet to either side of the pavement has been planted and the area looks much better. New oaks have been ordered to replace the ones lost during the hurricanes of 2004, and we escaped 2005 Hurricane Wilma with only tropical storm force winds. Things are looking up for Lakewood.

CHAPTER 9

HURRICANE WILMA, OCTOBER 12-25, 2005

Wilma formed and became an extremely intense hurricane over the northwestern Caribbean Sea. It had an all-time lowest central pressure for an Atlantic basin hurricane, and it devastated the northeastern Yucatan peninsula. Wilma also inflicted extensive damage over southern Florida.—Pasch, Richard J., Eric S. Blake, Hugh D. Cobb III and David P. Roberts, National Hurricane Center, 12 January 2006

The National Hurricane Center (NHC), in Miami describes the 2005 hurricane season as the most active on record with twenty-six named tropical storms, breaking the record of 21 set in 1933. Fourteen of these storms became hurricanes, breaking the previous record of 12 set in 1969. However, to the Florida citrus grower, Hurricane Wilma in late October, was the climax, the word climax being defined as the event or point of greatest intensity or interest. For the Florida grower it was the culmination of the 2005 season, although the overall monetary damage from Hurricanes Katrina and Rita exceeded the damage from Wilma. Wilma finished what hurricanes Charley, Frances and Jeanne in 2004 had only started in that she made a direct hit on the groves in Collier and Hendry counties which had escaped the hurricanes a year earlier. In so doing she contributed to the spread of Citrus Canker.

The Meteorology

Hurricane Wilma originated in a large area of disturbed weather stretched across the Caribbean Sea during the second week of October. As the western Caribbean is a primary source of tropical storms and hurricanes during the latter part of the hurricane season, this area was being closely monitored, and on Saturday October 15, 2005, the system was upgraded to Tropical Depression No. 24 and the first advisory was issued by the National Hurricane Center in Miami. The advisory indicated that the center was located near 17.6°N, 78.6°W and was moving slowly to the west at only about 3 mph. A request was issued for three hourly reports from ships within 300 miles of the center. The minimal central pressure was 1004 mb.

The National Hurricane Center continued to issue advisories at six hour intervals with intermediate advisories at three hour intervals. Advisory Number 5 at 5 P.M. on Sunday October 16 noted that aircraft data indicated that the depression had not strengthened. The center was located near latitude 17.7 north, longitude 79.6 west, 150 miles southeast of Grand Cayman Island. Steering currents were weak, and the depression was moving at only 2 mph. The minimum central pressure was 29.56 inches. Upgrading to tropical storm status was expected within the next 24 hours.

The expected upgrade was confirmed by Advisory Number 8 at 11:00 A.M. on Monday, October 17, 2005, and the storm was assigned the name Wilma. The center was located near 16.3 north, 80.0 west. It was moving to the southwest at about 5 mph with estimated central pressure at 29.47 inches.

Advisory Number 12 at 11:00 A.M. EDT on Tuesday, October 12 indicated that Wilma had become the twelfth hurricane of the season and was strengthening as she moved to the northwest at 7 mph as a category 1 hurricane on the Saffir-Simpson Scale. She was expected to become a major hurricane in the western Caribbean and for the first time it was noted that interests in the Florida Keys and the Florida peninsula should begin to monitor the progress of Wilma.

Hurricane Advisory Number 17 issued by the National Hurricane Center at 11:00 A.M. EDT on Wednesday, October 19 indicated that Wilma had become a potentially catastrophic category 5 hurricane as it moved to the west-northwest at 7 mph toward the Yucatan peninsula. Maximum sustained winds were near 125 mph with higher gusts. Appropriate watches and warnings were issued for Mexico, Swan Island, the Cayman Islands, Jamaica, Belize, Honduras, Nicaragua and Cuba.

At 11:00 A.M. on Thursday, October 20, 2005, the NHC issued a Discussion Number 21 which advised that while Wilma continued on a west-northwest heading she was expected to turn to the northwest within 12-24 hours, and due to a weakening of ridging in the Gulf of Mexico an acceleration to the northeast might occur in the future. An ominous statement noted that the main focus of the threat was from central Florida southward through the Keys, but that the hurricane was expected to weaken in the Gulf.

At 11:00 A.M. on Friday, October 21, 2005, Hurricane Wilma Discussion Number 25 advised that the center of Wilma was forecast to make landfall on the northeastern Yucatan. Its effect on Florida would depend on how long the hurricane center remained over land. Discussion Number 29 issued at 11:00 A.M. EDT,

Saturday, October 22, 2005, noted that Wilma remained just inland near Cancun, Mexico and would weaken slightly while it remained inland over the Yucatan, but there would be some opportunity for it to re-strengthen over the Gulf of Mexico. A hurricane watch was issued for all of the Florida Keys including Florida Bay, and it was expected that a hurricane watch would be issued for portions of central and southwest Florida.

Wilma Discussion Number 33 at 11:00 A.M., Sunday, October 23, 2005, showed Wilma to be increasing in speed. The best estimate of landfall intensity on the Florida peninsula was category 2 on the Saffir-Simpson Scale but an increase to category 3 was possible. This was confirmed by Hurricane Wilma Advisory Number 35 at 11:00 P.M. which announced that Wilma had strengthened to category 3 status. The eye was located near 24.4 north, 83.7 west and was moving to the northeast at 18 mph. Landfall was expected along the southwest Florida coast early the morning of Monday, October 24 with maximum sustained winds of 115 mph and higher gusts. The minimum central pressure was reported to be 28.29 inches. The hurricane warning included the Florida west coast from Longboat Key south, and along the Florida east coast from Titusville southward, and included Lake Okeechobee. Scattered tornados were expected and it was predicted that Wilma would be a category 2 hurricane even after crossing peninsula Florida to the east coast.

Hurricane Wilma Advisory Number 36 at 5:00 A.M. EDT, Monday, October 24, 2005, led with the statement that Wilma was stronger and approaching the southwest coast of Florida. The hurricane center was located near 25.5 north, 82.4 west, about 55 miles southwest of Naples, Florida. Forward speed was near 20 mph and expected to increase. Wilma was a large hurricane and south Florida was expected to receive hurricane force winds well before the arrival of the center. Data from the air force hurricane hunter plane indicated that Wilma was a category 3 hurricane with maximum sustained winds near 125 mph with higher gusts. Estimated minimum central pressure was 28.05 inches. Rainfall four to six inches was expected, and storm surge flooding of five to nine feet above normal along the Keys and Florida bay as well as Lake Okeechobee.

Intermediate Advisory Number 36A issued at 7:00 A.M. on October 24 noted that Wilma was moving into southwestern Florida and warned that tornados were possible over the central and southern Florida peninsula. Hurricane force winds extended outward up to 90 miles from the center. A wind gust to 95 mph was reported at Everglades City. Intermediate Advisory 36B reported Wilma to be centered over southern Florida near 26.3 north, 80.7 west or about 45 miles southwest of West Palm Beach, Florida. Sustained winds were near 110 mph, category 2 on the Saffir-Simpson Scale. Opa-Locka airport reported sustained winds of 85 mph with a gust to 105 mph. Minimum central pressure had risen slightly to 28.11 inches. It warned that tornados were still possible.

Hurricane Wilma Advisory Number 37 issued at 11:00 A.M. EDT warned against hurricane force winds from Longboat Key southwest on the west coast and Titusville southward on the east coast and including Lake Okeechobee. The center was located at 26.9 north, 80.0 west or about 15 miles north-northeast of West Palm Beach, Florida moving to the northeast at 25 mph. Maximum sustained winds were 105 mph with higher gusts. Hurricane force winds extended outward up to 100 miles from the center. Minimum central pressure was estimated at 28.23 inches.

Special Hurricane Wilma Advisory Number 38 at 2:30 P.M. EDT noted that Wilma had regained category 3 status off the Florida coast. Hurricane warnings along the Florida coasts and Lake Okeechobee were changed to Tropical Storm warnings, and at 5:00 P.M. Advisory Number 39 noted that Wilma was moving away from Florida at a speed of 37 mph. Wind speeds and rainfall at key locations are shown in Table 9-1.

Chet Townsend Reports

As he did for the 2004 hurricanes, Chet Townsend kept his UltimateCitrus.com, Citrus News readers advised of Wilma's movements as he rode out the storm at his home at Fort Denaud in Hendry County, midway between LaBelle and Alva. Chet's *Special Hurricane Edition—Report #1* on Monday October 24 stated:

Hurricane Wilma made landfall on Florida's southwest coast near Marco Island, Florida at around 6:00 A.M. EST. At the time of landfall, Wilma was a category 3 storm with winds of 125 mph.

At my house, the winds increased throughout the night with rain squalls beginning around midnight. As of 5:45 A.M., I have had 1.82 inches of rain. Currently the winds are around 25 mph gusting to around 40 mph. Currently, I still have electricity and a working Internet connection. I will continue with periodic updates throughout today, so stay turned to your email box.

Special Hurricane Edition—Report #2

Hurricane Wilma is still a category 3 storm with winds of 125 mph. The eye is currently southeast of Immokalee, Florida.

The cumulative rainfall at my house is now 3.18 inches of rain, as of 7:30 A.M. EST. On top of the 3.60 inches received on Wednesday, there is standing water in my Valencia grove. Currently the winds are around 40 mph gusting to around

TABLE 9-1. WIND SPEEDS AND RAINFALL FROM HURRICANE WILMA.

Location	Sustained Winds (mph)	Peak Gusts (mph)	Rainfall (in.)
Belle Glade	78	117	—
Clewiston	61	84	—
Fort Lauderdale	70	99	3.04
Fort Myers	59	79	5.44
Islamorada	—	108	—
Lake Okeechobee—SWFMD (L006)	92	112	7.00
Lakeland	—	—	7.34
Loxahatchee	71	113	3.12
NASA LC39B	64	94	2.79
Naples Pier	84	102	—
Opa Locka	85	105	—
Orlando (MCO)	40	48	5.17
Patrick AFB	—	74	6.95
Pompano Beach	83	98	—
Port Canaveral USCG	52	79	—
Punta Gorda	52	70	3.93
St. Petersburg	38	49	1.64
Tampa	35	44	1.04
Titusville Coop	—	—	6.90
Venice	—	—	7.45
Vero Beach	40	55	5.53
West Palm Beach	82	101	1.07
Winter Haven	36	46	4.77
	Unofficial Observations		
Arcadia (FAWN)	39	64	—
Boynton Beach (West)	90	119	—
Collier County EOC	—	125	—
Dry Tortugas Nat'l Park	—	133	—
Everglades City	—	135	—
Fort Pierce (FAWN)		—	6.09
Lake Alfred (FAWN)	—	44	—
Mountain Lake		—	6.00
Ona (FAWN)	23	53	—
Palm Beach (Missile Trading Annex)	82	114	—
Pompano Beach	—	120	—
Sebring (FAWN)	24	60	—
South Fork St. Lucie River	—	133	—
Stuart Skywarn Spotter	—	108	—

Source: Richard J. Pasch, Eric S. Blake, Hugh D. Cobb III and David P. Roberts. National Hurricane Center, January 12, 2006.

60 mph coming out of the east-northeast. It is now dawn, but I cannot see any fruit on the ground yet.

Currently, I still have electricity (although it has blinked on and off many times) and a working Internet connection. I had a report that parts of Ft. Pierce have already lost electricity far from the current storm center. Parts of Naples, Ft. Myers and Immokalee have also lost power. I will continue with periodic updates throughout today.

Special Hurricane Edition—Report #3

At my house west of LaBelle, we are getting the worst winds now as the strongest weather is northwest of the eye wall of Hurricane Wilma. The wind at my house has shifted and is now coming out of the north-northwest and the temperatures have dropped into the upper 60s. It is probably sustained around 75 mph with gusts over 100 mph. According to radar, the center of the very large eye of Hurricane Wilma is about 30 miles to the southeast of Clewiston. Most of Hendry County and northern Collier County near Immokalee are getting hammered right now with similar winds and rain as I am experiencing at the house as the worst weather is located to the northwest and west of the eye.

I did lose electric power just after 8:00 A.M. shortly after my last update. My laptop is running on battery as it is too dangerous to start up the generator yet.

The cumulative rainfall at my house is probably now about 5.00 inches. Due to the power outage, my automated weather station is not reporting through the receiver, so I don't have an actual update.

The weather is still too dangerous to venture out of the house to check the Valencia orange grove. But from my front door, I can still see fruit and leaves on the trees. I can not see any fruit on the ground due to the distance. However, the one Laurel or Water Oak tree in my yard has lost about 75% of its leaves. The

Live Oaks have lost about a third of their leaves so far. There are plenty of tree limbs and palm fronds on the ground that will keep me busy for at least a full day cleaning up.

I will continue with periodic updates through today. Please email me any reports you wish to share with Citrus News readers.

Special Hurricane Edition—Report #4

Well, as promised, Hurricane Wilma passed through southwest Florida very quickly. It is now 11:30 A.M. EST and the rain has stopped and the winds have greatly decreased, to about 20-25 mph. Enough that I was able to get out and take some pictures of my Valencia grove and my neighbor's hamlin grove.

My very preliminary damage report from Fort Denaud, Florida (about half way between LaBelle and Alva, FL) is that hamlin trees had the most fruit drop, estimated between 10-15%. Valencias fared better, estimated fruit loss of 5-10%. Some of my neighbors large hamlin trees and resets were bent over. I will continue with periodic updates throughout the day. Please email me any reports you wish to share with Citrus News readers.

Special Hurricane Edition—Report #5

By 1:00 P.M. EST, the sun had come back out in southwest Florida. There is a very nice breeze and the dew points have dropped with the arrival of the cold front. Temperatures are only in the mid 70s. Much nicer than after Charley, Frances, and Jeanne, not to mention Katrina and Rita, where the temperatures were in the high 90s.

Shortly after my last update, the battery on my laptop died, so I had to fire up the generator to charge it up again. My generators are going, powering my fridge, freezer, TV, etc., plus powering my well pump, so we have running water again. My house came through fine, but there are a lot of limbs on the ground. Please email me any reports you wish to share with Citrus News readers. This will be the last update today, unless I receive some breaking citrus related news.

November Report

The November report from the Florida Field Office, National Agricultural Statistics Service, dated November 10, 2005, described Hurricane Wilma as follows:

"Hurricane Wilma entered the State Monday the 24th at about 6:30 AM south of Naples as a category 3 storm with winds at 125 mph. The path took it over the lower portion of the Gulf citrus production area with hurricane force winds reported 60 to 80 miles on each side. It moved quickly without losing strength and passed about 30 miles south of Lake Okeechobee leaving the State over Palm Beach. As it left the State, the trailing eye wall produced winds as high as the front wall and exceeded 100 mph in citrus producing counties on the East Coast north to Indian River.

The hurricane was steered southward by a strong cold front approaching from the northwest. The cold front brought near record low temperatures for this time of year. There was no rain the rest of the week. Damage has been reported as severe to moderate in areas north to lower Highlands County and on the East Coast north to Indian River County. The heaviest loss of fruit is reported closest to the eye of the storm. Tree loss is reported moderate to light, mostly older or diseased trees. Leaf loss is reported light. Varieties which lost the most fruit were grapefruit, Navel and hamlin oranges and early tangerines. Harvest was interrupted as growers and others cleared roads and recovered equipment affected by the storm. Surface water levels are very high, growers are working to reduce canal levels and dry out the groves."

Michael S. (Mike) Murphy is Chief Executive Officer and General Manager for Cooperative Producers, Inc., Cooperative Three, Inc., Gulf Harvesting, Inc. and Ranch One Cooperative, Inc. with offices in Immokalee, Florida. Mike kindly provided the following observations on the effect of Hurricane Wilma on growers in Collier, Hendry and Lee counties.

The Cooperative groves are comprised of three separate citrus plantings totaling approximately 7,600 grove acres located in Collier, Hendry and Lee counties. The first planting is the Cooperative Producers Incorporated grove (CPI) located in southwest Hendry County with over 3,600 grove acres of various citrus varieties planted in 1985-86. The next, planted in 1987-88, is the Ranch One Cooperative grove (ROC). This grove is over 3,200 grove acres of various citrus varieties and located in the heart of Collier County. The last project is the Cooperative Three Incorporated grove (CTI) located in southeast Lee County with 700 grove acres of various citrus varieties and was planted in 1991-92.

The ROC grove was the closest to Wilma's eye wall, we estimate within 12 miles, and received the most crop and tree damage. The wind speed was recorded at over 135 mph when Wilma slammed the coast some 35 miles south of the ROC. We estimated 80% of the Sunburst crop (25,000 boxes), 45% of the hamlin and pineapple crop (60,000 boxes), with approximately 3500 boxes salvaged, and 35% of the Valencia crop (251,000 boxes) blown to the ground based upon actual pick-outs to date. The droppage has been steady since the storm passed, with peak drop amounts 2-3 weeks immediately after Wilma and continuing until harvest.

Tree damage was extensive but randomly scattered throughout the grove and appeared to be consistent with weight, size and maturity of the crop. For example, we experienced more crop, limb and tree loss in our early and midseason varieties than in our Sunburst and Valencia trees. The worst tree damage occurred on the northern bound of all blocks, with the exception of sporadic areas affected by suspected vortex action. These areas reflected trees blown over from all directions. It is clear to us, as we went into a block, each tree acted as a buffer to the next; thereby, lessening the tree and crop damage. This is the result of a north to south tree row planting.

To date we have repositioned (stood back upright) over 3,200 trees, with over 98% survival rate. We are still counting the total trees blown over, but 5,000 (2-3%) is a good estimate.

Rainfall amounts received over a 2-day period were between 10 and 12 inches. We were able to evacuate the water, to a controllable stage, in a period of five days. Little tree damage was sustained due to water.

The CTI grove experienced the most fruit loss per acre due to the high percentage of grapefruit acreage. It is estimated that 55% of the crop (88,000 boxes) hit the ground, but salvage began immediately and the actual estimated loss was at 42%. On the other hand, we were able to pack our Fallglo crop that was virtually unscathed due to the protection of the block from windbreaks created by wetlands.

Tree and limb damage was limited due to the distance from the eye, estimated at 40 miles. Approximately 100 trees were blown over.

Rainfall amounts over a 2-day period were between six and eight inches. The drainage system was taxed excessively, but no breaching of the reservoir was reported. Water was evacuated in approximately eight days. Tree stress was present due to extended periods of high water levels in the swales.

The CPI grove was approximately 35 miles from the eye wall and received most of the damaging wind from the north. The highest sustained winds, as reported at Ft. Myers, were over 90 mph. The CPI grove received extensive tree, limb and crop damage in some areas, but much like the ROC grove was random or on the northern bounds of the blocks. The degree of crop damage was determined by variety, size, maturity and weight. We had several areas where vortex activity was suspected due to the positions of downed trees. We estimated over 4,000 trees were blown over by the storm, with close to 3,500 of those repositioned. As with the ROC grove, time will tell, but 98% survival rate is expected.

The crop loss at the CPI grove was less than at the ROC grove, but still substantial. Hamlin and mid-season varieties were estimated at 31% (175,000 boxes) blown to the ground, with approximately 8,000 boxes salvaged. The grapefruit crop was estimated at 50% (75,000 boxes) blown to the ground, but salvage operations reduced the loss to approximately 15,000 boxes. The Valencia crop was estimated at over a 20% loss (150,000 boxes).

Rainfall amounts were between six and eight inches over the 2-day period. Water was evacuated off the property in four days and was not a factor in tree damage or in the salvage operation.

Within hours of Wilma's passing and tropical storm winds still pounding the groves, employees were assessing the damaged caused by the storm. Having our own harvesting company crews and the cooperative relationship with Florida's Natural Growers, played an important role in lessening the crop loss. The salvage operation began the day after Wilma passed and continued until spoilage and financial feasibility was no longer practical. On the first day, nine loads were harvested, with 20 loads harvested on the second day and upward from there. Over 70,000 boxes were utilized within the first two weeks of Wilma. General Statement—all groves have responded well to increased nutritional sprays applied to replenish the elements depleted by Wilma and are showing signs of a better than average crop. Time will tell on the 7,000 trees we set back up, some were set back as late as February. All in all, the Cooperative groves have been blessed. Now if we could just get a little rain, the fall's weather will be nothing but a memory.

George Hamner, Jr. is President of Indian River Exchange Packers in Vero Beach. George recalls that there was very little damage to infrastructure from Wilma, which came in the back door across the Everglades. However, the loss to the fruit crop was significant. George estimated 25-30% loss of grapefruit in Indian River County, a 50-60% loss in St. Lucie County and 80% in Martin County. Before Hurricane Wilma, it was estimated that the state had 24 million boxes of grapefruit of which 16 million were in the Indian River district. After Wilma the state estimate had dropped to 16 million boxes, 8 million of which were in the Indian River district.

Gregory P. Nelson is President of DNE World Fruit Sales in Fort Pierce, Florida. DNE is a major grower and packer of grapefruit, oranges and specialty fruit from its own groves and from other packinghouses. DNE groves are principally located in Indian River, St. Lucie and Martin Counties on the Florida east coast and Hendry County in southwest Florida.

Mr. Nelson indicated that DNE suffered significant damage from Hurricane Wilma in their Hendry County groves which escaped damage from the three hurricanes in 2004. While not big on grapefruit in Hendry County, only 300 acres, 70% of the fruit was on the ground. Sixty-five percent of the oranges in Hendry County were on the ground.

Jeff Cusson is President and **R. Scott Hurley** is Director of Fresh Fruit for Becker Holding Company of Ft. Pierce. Scott remembers that the eye of Wilma came directly over Becker's Martin County groves on its way back out and over the Atlantic Ocean, the opposite direction of Hurricanes Frances and Jeanne a year earlier. It was a "meaner storm" than had been forecast, probably because it stayed over water as it crossed the Everglades. Being late in the season, the fruit was heavier due to its higher juice content which made it worse than Frances and Jeanne. Grapefruit losses totaled 90% in Martin County, 70% in St. Lucie County and 50% in Indian River County (see Figures 9: 1-4).

Mr. Nat Roberts, Managing Partner and **Mr. Mark DuBois**, Production Manager of Callery Judge grove in Palm Beach County describe major damage both to the fruit crop and to barns and pump stations when the west wall of Hurricane Wilma passed over the almost 4,000 acre grove. According to Roberts, the center of Wilma "passed right over the top" with the west eyewall being much more destructive than the east eyewall. Wind speeds in the east eyewall were estimated at 85-90 mph causing some damage, but when the west eyewall passed the wind speeds were higher and the damage much

Figures 9: 1-4 illustrate the magnitude of grapefruit losses in flooded St. Lucie county groves after Hurricane Wilma. Source: Jeff Cusson, Becker Holding Company.

more severe. Three barns were totally destroyed as were the covers of all seven pump stations and 80% of the fruit hit the ground.

Roberts indicated that of all varieties, Fallglo did best. He attributed this to the "willowy" tree which would swing back and forth with the wind. Most of the Fallglo crop was sold after Wilma, while the heavier grapefruit was a 90% loss. Murcotts were generally a 40-50% loss, but some blocks suffered as much as 80% lost. The storm left nothing but a few shiners in a 150 acre block of 35 year old Murcotts.

After Frances and Jeanne in 2004 and Wilma in 2005, the trees had been stripped two years in a row. As a result, the 2006-07 crop is questionable.

From Peter Spyke, Ft. Pierce, Florida

Peter Spyke can always be depended on to share his observations as follows:

Well, we made it. I have the generator running, and we still have phone lines, so I'm able to get out on e-mail.

It blew hard. I've never seen anything with as much power. I was not in St. Lucie County during Frances or Jeanne, so I can't really compare.

There is a lot of fruit on the ground. Around the house, almost all the Navels are on the ground. I rode around a bit right before sunset and saw grapefruit groves with probably 30-50% of the fruit on the ground. It was not a representative sample, though. I think it's safe to say that we lost a significant amount of grapefruit. I can't say about the Valencias, etc.

My pressure tank blew over and broke all the plumbing, so my afternoon was mostly concerned with fixing that. I was glad we're an irrigation company—had lots of fittings and pipe, so it wasn't too bad.

A lot of trees are stripped of leaves and fruit. Healthy groves that were densely foliated fared better, of course. Weather is cool, beautiful sunset. Hard to believe that only a few hours ago it was so bad.

Horticultural Research Laboratory

Dr. Calvin E. Arnold is Director at the U.S. Horticultural Research Laboratory in Ft. Pierce, Florida. In comparing Hurricane Wilma in 2005 to Hurricane Frances and Jeanne in 2004, Dr. Arnold indicates that while roof damage at the laboratory was significant after Frances and Jeanne, it was not noteworthy after Wilma. However, there was additional greenhouse damage after Wilma.

Hurricanes Frances and Jeanne had damaged all 36 of the research greenhouses to some extent, and required $1.2 million dollars for repair and restoration. Fortunately we received the funds within a week after the hurricanes, and repairs were completed four months before we were struck by Hurricane Wilma who came in the back door rather than off the ocean. Wilma resulted in an additional $180,000 damage to the greenhouses, but in this case the funds will have to come out of our regular budget so repairs have not been completed as of today (January 15, 2006).

Frances and Jeanne had pushed over a lot of our trees which we had propped up prior to Wilma. Wilma pushed them over again but we did not actually lose a lot of trees. After all three hurricanes, there was only nominal loss of germplasm in the breeding program. However, the loss of fruit from the breeding program was a negative as without fruit to be evaluated we are a year behind.

Hurricane Damage to University of Florida Laboratories

The combination of damage to University of Florida laboratories by the 2004 hurricanes, followed by Hurricane Wilma in 2005, is covered extensively by an article entitled "Labs Blown for a Loss" by Susan Salisbury in the Palm Beach Post dated Jan. 22, 2006.

Quoting University of Florida sources, Salisbury describes damage at five different research laboratories of the Institute of Food and Agricultural Sciences (IFAS), as well as the Florida Medical Entomology Lab, and summarizes needs and costs to upgrade each facility to its pre-hurricane condition. Details from the Salisbury article are as follows:

Florida Medical Entomology Laboratory, Vero Beach: $375,000 to replace storage sheds.

Indian River Research and Education Center, Fort Pierce: $675,000 to upgrade to hurricane standards and repair greenhouses.

Everglades Research and Education Center, Belle Glade: $1.1 million to upgrade laboratories.

Fort Lauderdale Research and Education Center: $2.6 million to replace air conditioning, heating.

Southwest Florida Research and Education Center, Immokalee: $3.8 million to replace/remodel four buildings, replace dated trailers and add labs and offices.

Tropical Research and Education Center, Homestead: $9.5 million for laboratory and support space for faculty members, three greenhouses and a conference room/classroom.

At the Belle Glade Center, Director **Dr. Christine Waddill** indicates that the most important losses were to ongoing research which had been in progress for many years. Twelve greenhouses were torn down and 55 of the Center's 60 buildings severely damaged, eight of which will need to be demolished. Waddill also noted that because of the age of so many of the buildings at the Belle Glade Center which dated to 1921, and at the Tropical Research and Education Center at Homestead dated to 1929, the Federal Emergency Management Center (FEMA) would only provide 15% of the costs to rebuild.

Dr. Brian Scully, Director of the Indian River Research and Education Center at Fort Pierce, added that Hurricane Frances in 2004 destroyed the Aquaculture Greenhouse. It was rebuilt only to be destroyed again by Hurricane Wilma in 2005. He further noted that "As we reconstruct, I am under the assumption that we will be hit by hurricanes every year."

Southwest Florida Research and Education Center, Immokalee, Florida

Dr. Bob Rouse is a Citrus Horticulturist at the University of Florida, Southwest Florida Research and Education Center in Immokalee, Florida. In an interview with the author, Dr. Rouse relates that Hurricane Wilma did major damage to the Immokalee Center except for the main building which was untouched. The modular buildings and five greenhouses were taken apart and will have to be replaced.

All of the fall vegetables were destroyed except for one tomato field. The citrus grove came through fairly well except that the swales were full of water. Fortunately we were able to remove the water within 72 hours to avoid root damage. About a dozen trees were blown over and another 15-20 had broken scaffolds but could be salvaged.

The structure of the two Budwood screenhouses held up fine, but the poly roof was half torn and blown off. The roof was quickly restored to protect the integrity of the screenhouses. Other damage was patched or replaced.

Grapefruit loss in the grove was major at 90%. Hamlins were also hit hard with 65% loss but only 20-25% of the Valencias. Obviously this was due to the size and weight of the different varieties early in the season. Grapefruit was heaviest, then hamlin oranges, then Valencias.

Effect of Wilma on the Estimated Crop for the 2005-06 Season

When the estimate for the 2005-06 crop was released by the Florida Field Office of the USDA National Agricultural Statistics Service (NASS) on October 10, 2005, it stimulated a note of optimism among Florida growers battered by Hurricanes Charley, Frances and Jeanne. The initial estimate for all oranges was 190 million boxes compared to a harvest of 149.6 million boxes for the 2004-05 season, and all grapefruit showed an increase from 12.8 million boxes in 2004-05 to a projected 24 million boxes in 2005-06. The tangerine estimate indicated an increase from 4.45 million boxes in 2004-05 to 6 million boxes in 2005-06. Temples were forecast to increase from 600,000 to 900,000 boxes, but tangelos, which escaped a direct hit in 2004-05 were expected to decrease from 1.55 million to 1.4 million boxes.

Then came Wilma. On October 31, 2005, Florida Citrus Mutual released preliminary figures representing an estimated statewide citrus crop loss impact due to Hurricane Wilma. Mutual's estimates indicated that the storm reduced the overall citrus harvest for the 2005-06 season by 17% of the total corp, approximately 35.7 million boxes of fruit. Based on price expectations following the October 12 U.S. Department of Agriculture crop estimate, this would result in a crop loss approaching $180 million for fruit only.

This figure did not represent tree loss and replanting costs or damage costs associated with barns, equipment, processing and packing facilities, which would increase the overall economic loss to the industry. In addition, given experience from the 2004 hurricanes, growers can expect more fruit to fall within the next few weeks, further adding to the overall fruit loss.

"This is a very preliminary figure that could increase as additional damage is discovered and industry losses are calculated," said Andy LaVigne, Florida Citrus Mutual's Executive Vice President/CEO.

For oranges, estimates show a preliminary loss of 24.4 million boxes, which represents 13% of the state's orange crop. For grapefruit, estimates show a preliminary loss of 11.3 million boxes, which represents 47% of the state's grapefruit crop. Later when NASS released its updated forecast on December 9, 2005, any previous optimism had been replaced by pessimism. Not only was there a major fruit loss, but there was heavy damage in Hendry and Collier counties which had escaped the major devastation in 2004 and had previously been spared the spread of Citrus Canker. The southwest counties now appeared certain to be riddled by canker spores blown in by Wilma.

Effect of Wilma on the Fresh Fruit Industry

In Chapter 7, figures were presented showing the decrease in fresh fruit shipments after Hurricanes Charley, Frances and Jeanne. The following figures (Table 9-2) compare fresh fruit shipments by variety, after Hurricane Wilma, with shipments after the 2004 hurricanes. The numbers are as of April 30, in each case, and do not represent a full season. Our printing deadline prevented complete figures through July 31, 2006. The figures in parentheses represent the percentage of total shipments of the variety.

TABLE 9-2. COMPARISON OF FRESH FRUIT SHIPMENTS BY VARIETY.

Variety	Thru 4-30-05	Thru 4-30-06
	4/5 Bushel boxes	
Grapefruit	13,199,941 (46%)	12,222,033 (45%)
Oranges	8,354,672 (29%)	7,642,115 (28%)
Tangelos	628,019 (3%)	658,186 (2%)
Temples	313,033 (1%)	284,051 (1%)
Early Tangerines	3,261,978 (11%)	3,395,992 (13%)
Murcotts	2,219,829 (10%)	2,869,965 (11%)
Total	28,477,475	27,114,338

Source: Citrus Administrative Committee, Fresh Shippers Report, 2006.

Everglades City

Chokoloskee and Everglades National Park

Charles "Charlie" Hendrix is a 50-year veteran of the citrus processing industry. On November 7 and 8, 2005, two weeks after the eye of Hurricane Wilma made landfall in Collier County, Charlie and his wife Margaret made a visit to Everglades City, Chokoloskee and the Everglades National Park area, and reported in a letter to the author as follows.

Wilma came ashore at 6:30 A.M. near Cape Romane and crossed the peninsula from west to east at 20 mph. Most of the damage seems to have occurred on the backside of the hurricane after the eye had passed. On top of high tide the flood waters in Chokoloskee Island were near 9 feet and in Everglades City 4 feet. The destruction was beyond belief. Two weeks after the hurricane, debris was still being hauled out to the highway for disposal. The debris contained TV sets, refrigerators, washing machines, dryers and even automobiles.

We visited the ranger station and found that repairs were being made and cabins would be available later in the week. Heavy boarding had saved the office building though shingles were lost from the roof. Electricity had been restored.

Four campsites in the park, Kingston Key, Broad River, Sunday Bay and Sweetwater were closed as everything had been washed away. Charlie feared that these sites might be lost for good, but in a conversation with the ranger station on February 14, 2006, he found that the Sunday Bay and Sweetwater sites had been restored and Kingston Key was undergoing repairs. However, the Broad River site would no longer be at the same location. A request has been made to build a Chickee campsite nearby that would be easier to maintain.

Chokoloskee is a small island with a year-round population of only 200, but as many as 1,000 during the winter months. Many residents live in mobile homes and lost furnishing, carpets and appliances, but no one was killed despite the fact that the residents rode out the storm without evacuating. It was the most damaged community in Collier County.

Hurricane Assistance Programs

A special addition of the Farm Service Agency, Florida State Newsletter for May 2006, reported that the USDA has four programs for assistance to growers damaged by 2005 hurricanes. Sign-up began May 17, 2006 for a Hurricane Indemnity Program (HIP), A Tree Indemnity Program (TIP), a Fed Indemnity Program (FIP) and a Livestock Indemnity Program (LIP).

Eligible Counties

Only those counties that received a Presidential or Secretarial Disaster Designation due to the calendar year 2005 hurricanes were eligible for these four programs. There were 36 eligible counties in Florida: Bay, Brevard, Broward, Calhoun, Charlotte, Collier, Dixie, Escambia, Franklin, Gadsden, Glades, Gulf, Hardee, Hendry, Highlands, Holmes, Indian River, Jackson, Jefferson, Lee, Leon, Levy, Liberty, Martin, Miami-Dade, Monroe, Okaloosa, Okeechobee, Palm Beach, St. Lucie, Santa Rosa, Sarasota, Taylor, Wakulla, Walton and Washington.

Program Sign-up

The sign-up for HIP, TIP, FIP and LIP began May 17, 2006. Applications MUST be filed in the County Office where the loss physically occurred.

Payment Limitations

No "person" may receive more than $80,000 in each program—HIP, TIP, FIP and LIP. An individual or entity whose average adjusted gross income (AGI) exceeds $2.5 million is ineligible for HIP, TIP, FIP and LIP benefits unless 75 percent or more of their income is derived from farming, ranching and/or forestry.

The programs of greatest interest to citrus growers affected by Hurricane Wilma were the Hurricane Indemnity Program (HIP) and the Tree Indemnity Pro-

TABLE 9-3. TOP FRESH FRUIT SHIPPERS THROUGH APRIL 30, 2006.

Shipper	Total 4/5 Bushel Boxes
Dundee Citrus Growers Association	3,011,626
Haines City Citrus Growers Association	1,505,862
Bernard Egan and Company	1,417,266
Greene River Marketing	1,329,899
Duda Farm Fresh Foods, Inc.	1,218,663
Lake Placid Growers, LLC	1,181,311
Packers of Indian River Ltd.	1,162,048
Sun Ag., Inc.	1,134,906
Emerald Packing Company	884,205
Ben Hill Griffin, Inc.	879,703
Heller Brothers Packing Corp.	843,884
Minton Sun, Inc.	825,234
W. G. Roe and Sons	741,966
East Coast Packers, Inc.	734,901
Hunt Brothers Coop.	682,647
Premier Citrus Packers, LLC.	681,380
Leroy Smith, Inc.	669,743
Riverfront Packing Company, LLC	598,231
IMG Citrus, Inc.	563,908
Rio Citrus	559,533

Source: Citrus Administrative Committee, Fresh Shippers Report, 2006.

gram (TIP). The HIP provided assistance to eligible producers who received either a FCIC crop insurance indemnity payment or a crop loss payment under NAP for losses that were attributable to the 2005 hurricanes. HIP payments were equal to 30 percent of the crop insurance indemnity or 30 percent of the NAP payment not to exceed 95 percent CAP.

Producers must be able to certify that the production loss was wholly or partially because of an eligible cause of loss and was the direct result of one or more of the eligible hurricanes. Eligible causes of loss are those directly related to the hurricanes and will include the following:

NAP and FCIC Crops

Excessive moisture, precipitation, and/or rain, flood, excessive wind, cyclone, tornado, hurricane or tropical depression.

FCIC Crops Only

Storm surge; or salinity because of salt water intrusion.

The commodity must have been located in an eligible county and suffered a loss caused by an eligible disaster during the disaster period. Producers must have been paid a 2005 or 2006 FCIC crop insurance indemnity payment or a 2005 or 2006 payment under NAP for eligible production losses.

The TIP provided assistance to eligible producers who suffered fruit tree, bush, and vine losses attributable to the hurricanes. Compensation was based on the geographic location of the stand associated to the established tiers. TIP benefits were related to the cost of replanting, rehabilitation, cleanup, and debris removal. **Timber, nursery inventory, and container grown fruit trees, bush, and vine losses were not covered.** Producers were not required to replant, rehabilitate, remove debris, or cleanup to receive TIP benefits.

To qualify for TIP, an eligible fruit tree, bush, and/or vine producer is one who bears financial responsibility and whose stand has incurred damage of at least $90 per acre for replanting, rehabilitation, cleanup, or debris removal, excluding crop production.

Eligible stands must be physically located in an eligible county; must have been impacted by an eligible cause of loss during an eligible disaster period; must be fruit bearing; and must have suffered at least $90 per acre damage.

This table lists payment rates per acre for each tier. The producer shall certify to the applicable payment level.

I	$750
II	$300
III	$200
IV	$90

The tiers represent the bands of damage generally correlating to the maximum sustained winds of the applicable hurricane. Each tier was established at the national level and maps were provided to each County Office.

To apply for TIP benefits, eligible producers shall file an automated FSA-573 in the county where the claimed fruit tree, bush, and vine losses physically occurred during the applicable disaster period. Each eligible producer with a share in the stand must complete and sign a separate FSA-573 for their share of the benefits.

All applicants' signatures had to be received by the ending sign-up date. There were **no** late-file FSA-573 approvals.

Wilma's Brothers and Sisters

While this book and chapter are devoted to hurricanes affecting the Florida citrus industry, the 2005 Atlantic Basin hurricane season is of great interest because of its record breaking destructive hurricanes and its record numbers of tropical storms and hurricanes.

The Southern Region News of the National Weather Service notes that the 2005 Atlantic Hurricane Season produced the following first time events: (1) at one time, Hurricane Wilma was the strongest hurricane ever recorded in the Atlantic Basin; (2) a record seven named storms formed by the end of July, including five which formed during July; (3) the 2005 Hurricane Season was the first on record to produce three category 5 hurricanes; and (4) 2005 was the first season in recorded history to **exceed** 21 named storms.

The December 1, 2005 Monthly Tropical Weather Summary from the National Hurricane Center in Miami states that the 2005 Atlantic Hurricane Season was the most active on record with 26 named tropical storms breaking the old record of 21 set in 1933. Fourteen storms including Cindy became hurricanes breaking the old record of 12 set in 1969. Seven of the hurricanes became major hurricanes, category 3 or higher on the Saffir-Simpson Scale including three, Katrina, Rita and Wilma which reached category 5 intensity. In addition, Hurricane Emily may have also briefly reached category 5. The season also included three depressions that did not reach tropical storm strength.

A December 1 Summary Table from the National Hurricane Center, prepared by forecasters Stewart, Beven, Avila, Franklin, Knabb and Pasch, follows as Table 9-4.

TABLE 9-4. A SUMMARY OF 2005 ATLANTIC BASIN TROPICAL STORMS (TS), TROPICAL DEPRESSIONS (TD) AND HURRICANES (H) THROUGH NOVEMBER.

	Name	Dates	Max wind mph	Deaths	U.S. Damage $ Million
TS	Arlene	8-13 JUN	70	1	Minor
TS	Bret	28-29 JUN	40	1	0
TS	Cindy	3-7 JUL	70	1	Minor
H	Dennis	4-13 JUL	150	41	1840
H	Emily	11-21 JUL	155	5	Minor
TS	Franklin	21-29 JUL	70	0	0
TS	Gert	23-25 JUL	45	0	0
TS	Harvey	2-8 AUG	65	0	0
H	Irene	4-18 AUG	105	0	0
TD	Ten	13-14 AUG	35	0	0
TS	Jose	22-23 AUG	50	6	0
H	Katrina	23-30 AUG	175	1200	80000
TS	Lee	28 AUG-1 SEP	40	0	0
H	Maria	1-10 SEP	115	0	0
H	Nate	5-10 SEP	90	0	0
H	Ophelia	6-17 SEP	85	1	1600
H	Philippe	17-24 SEP	80	0	0
H	Rita	18-26 SEP	175	6	9400
TD	Nineteen	30 SEP-2 OCT	30	0	0
H	Stan	1-5 OCT	80	100	0
TS	Tammy	5-6 OCT	50	0	Minor
STD	Twenty-Two	8-10 OCT	35	0	0
H	Vince	9-11 OCT	75	0	0
H	Wilma	15-25 OCT	175	22	14400
TS	Alpha	22-24 OCT	50	20	0
H	Beta	26-31 OCT	115	0	0
TS	Gamma	13-20 NOV	50	37	0
TS	Delta	22-28 NOV	70	0	0
TS	Epsilon	29 NOV-DEC?	70	0	0

Note: Dates based on coordinated universal time (UTC).
Source: National Hurricane Center, Miami, FL.

A brief synopsis of the 2005 storms in chronological order is as follows:

Tropical Storm Arlene—Formed June 8 near the northeastern coast of Honduras and crossed western Cuba with 50 mph winds. It's winds peaked at 70 mph in the eastern Gulf after which she made landfall near Pensacola, Florida with 60 mph winds. Damage was minimal.

Tropical Storm Brett—Became a tropical depression over the Bay of Campeche on June 28, strengthened to a tropical storm and made landfall near Tuxpan, Mexico with 40 mph winds on the 29th. Flooding was produced in the State of Veracruz, and there was one confirmed death.

Tropical Storm Cindy—Later reclassified as a hurricane, formed in the extreme western Caribbean Sea on July 3 east of the Yucatan/Belize border. She became a tropical storm on July 5 and made landfall near Grand Isle, Louisiana on July 6 as a minimal hurricane accompanied by heavy rainfall. Cindy caused one death in Georgia due to flooding. Otherwise, damages were minimal.

Hurricane Dennis—Was the first major hurricane of the season. It developed on July 4 from a tropical wave near the southern Windward Islands, strengthened to a tropical storm on July 5 and a hurricane on July 6. Dennis intensified into a major category 4 hurricane on July 8 with 150 mph winds, crossed Cuba with 145 mph winds, then weakened to a category 1 hurricane on July 9. Dennis reintensified to a category 4 hurricane on July 10, but weakened to the category 3 level before making landfall near Navarre Beach in western Florida late on July 10. Damage was estimated at $1.48 billion. Heavy rainfall occurred across Florida. There were 41 reported deaths, 22 in Haiti, 16 in Cuba and 3 in the United States.

Hurricane Emily—Was the second major storm of the season. Emily developed on July 11 from a tropical wave 1300 miles east-southeast of the lesser Antilles. She abruptly strengthened over the southern Windward Islands near Grenada on July 14 and became a major hurricane with 155 mph winds in the Caribbean south of Hispaniola. Passing south of Jamaica and the Cayman Islands, Emily made landfall over the Yucatan peninsula on July 18 and again over northeastern Mexico on July 20 with 125 mph winds. Five deaths were attributed to Emily.

Tropical Storm Franklin—A minor storm which formed July 21 in the central Bahamas. Franklin passed 185 miles west of Bermuda on July 26 and became extra-tropical on July 29. No deaths or casualties were reported.

Tropical Storm Gert—Another minor storm. It formed in the Bay of Campeche on July 23 and made landfall in Mexico on July 25. Rainfall was heavy.

Tropical Storm Harvey—A long-lived cyclone which formed from a tropical wave off the coast of Africa on July 23. It did not become a tropical depression until August 2 and a tropical storm until August 3. Meanwhile Harvey had passed through the northeastern Caribbean, and moved north to a point southeast of Bermuda where its winds peaked at 65 mph.

Hurricane Irene—A long lived storm which remained over the open Atlantic from August 7-18. For a time it was a category 2 hurricane north of Bermuda with 105 mph winds before becoming extra-tropical. No reports of damage were recorded.

Tropical Depression Ten—Developed from a tropical wave on August 13 about 1,050 miles east of Barbados. Strong wind shear caused it to dissipate the following day, but its remnants may have contributed to the initial development of Hurricane Katrina.

Tropical Storm Jose—Formed from a tropical wave in the Bay of Campeche on August 22 about 80 miles east-northeast of Veracruz. It strengthened to a tropical storm later the same day and made landfall about 35 miles north of Veracruz with winds near 50 mph. It was responsible for six deaths.

Hurricane Katrina—Probably the most devastating hurricane in U.S. history. She formed from a depression 175 miles southeast of Nassau, Bahamas on August 23 and became a tropical storm on the 24th. Moving first on a northwest course through the Bahamas she turned to the west toward south Florida and gradually strengthened to a category 1 hurricane and made landfall near the Dade/Broward County line late on August 25. Katrina then moved to the southwest across south Florida producing a foot of rain and damaging trees and power lines. After entering the Gulf of Mexico, Katrina strengthened to category 5 intensity on August 28 about 250 miles south-southeast of the mouth of the Mississippi River. Later that day Katrina's winds peaked at 175 mph and the pressure fell to 902 MB, the fourth lowest pressure on record. Turning to the northwest and then north, Katrina made landfall along the Louisiana/Mississippi border as a category 3 hurricane. Damage and casualties in the New Orleans area, along the Mississippi Gulf Coast and in south Florida were catastrophic. The death toll was the highest since the hurricane which struck Lake Okeechobee in 1928. The final death toll in the U.S. was estimated as over 1200 and damage was estimated at 80 billion dollars. Final calculations are still being tabulated at the time of this writing.

Tropical Storm Lee—First became a depression 925 miles east of the lesser Antilles on August 28, dissipated on the 29th but then redeveloped into a storm on the 31st, finally breaking up on September 1. There were no reports of damage or casualties.

Hurricane Maria—Began as a depression centered about 1100 miles east of the Windward Islands. Moving to the northwest she became a tropical storm on September 2 and a hurricane on September 6 about 470 miles east of Bermuda with 115 mph winds. Maria became extra-tropical on September 10 and finally reached the coast of Norway on the 14th where she caused a landslide and one death.

Hurricane Nate—Formed from a tropical wave off the coast of Africa on August 30, and increased to a tropical depression on September 5 about 30 miles southwest of Bermuda. Nate intensified to a tropical storm on the 6th and a hurricane on the 7th. After briefly threatening Bermuda, Nate passed well to the southeast of the island and weakened to a tropical storm on the 9th, becoming extra-tropical on September 10. There were no reports of damage or casualties.

Hurricane Ophelia—Became a tropical depression near Grand Bahama Island on the 6th and a tropical storm on the 7th about 115 miles east-southeast of Cape Canaveral. Ophelia moved erratically for two days then became a hurricane on September 8, after which she moved to the northeast and stalled off Cape Hatteras, North Carolina. Later she moved erratically off the Carolina coast with winds reaching 85 mph on September 14. Ophelia finally moved to the northeast and became extra-tropical near Nova Scotia on the 18th. Ophelia caused one death, a drowning off the southeast coast of Florida, and $1.6 billion in damage in the U.S. Significant beach erosion occurred from the central Florida coast to the North Carolina coast.

Hurricane Philippe—Formed from a depression which became a tropical wave and then a tropical storm on September 17 about 305 miles east of Barbados. Moving north-northeast Philippe became a hurricane 390 miles east of the Leeward Islands on the 19th with winds peaking at 80 mph on September 20, after which the winds diminished to tropical storm force. Philippe was absorbed into a non-tropical low on September 24. No reports on damage and casualties were received.

Hurricane Rita—Was a devastating storm comparable to Katrina and Wilma. She formed on September 17 from a depression east of the Turks and Caicos Islands and moved westward through the central Bahamas on the 19th as a tropical storm. Rita intensified rapidly to a category 2 level on September 20 while moving through the Florida Straits 50 miles south of Key West. While she did not make landfall, downed trees, storm tides and flooding occurred in the Keys. After entering the Gulf of Mexico, Rita intensified rapidly with winds of 165 mph on September 21 and 175 mph on the 22nd about 570 miles east-southeast of Galveston, Texas. The central pressure fell to 875 mb, the third lowest on record in the Atlantic Basin at the time. Rita produced storm surge flooding in portions of New Orleans before turning northwest and weakening to category 3 on September 23 and making landfall early on September 24 just east of the Texas/Louisiana border. Rita caused devastating storm surge flooding and wind damage in southwest Louisiana and southeast Texas. Six deaths and 9.4 billion dollars of damage were estimated.

Tropical Depression 19—Formed 665 miles west-southwest of the Cape Verde Islands. It moved slowly northeast and dissipated on October 2 causing no damage or casualties.

Hurricane Stan—Developed from a tropical depression 110 miles southwest of Cozumel, Mexico. The depression moved west-northwest and became a tropical storm as it reached the Yucatan peninsula on October 2. It decreased to a depression again, then regained tropical storm status on October 3 in the Bay of Campeche. Stan then moved to the southwest and became a hurricane on October 4 before making landfall 90 miles east-southeast of Veracruz, Mexico with 80 mph winds. Heavy rains resulted in flash floods and mud slides and contributed to 80 deaths.

Tropical Storm Tammy—Developed about 20 miles east of Cape Canaveral on October 5. It paralleled the Florida east coast for a day before making landfall near Mayport with 50 mph winds. Locally heavy rain occurred across southern Georgia and Alabama.

Tropical Depression 22—Formed about 450 miles southeast of Bermuda on October 8. It moved northwest to west and degenerated into low pressure area about 175 miles west-southwest of Bermuda.

Hurricane Vince—Developed from a low pressure area in the Atlantic which became a tropical storm on October 9 about 515 miles east-southeast of the Azores Islands. It became a hurricane later that day as it moved slowly past the Madeira Islands. It weakened to a tropical storm and then to a tropical depression before making landfall near Huelva, Spain. Vince was the first known tropical cyclone to make landfall in Spain.

Tropical Storm Alpha—As a result of extremely heavy activity during the 2005 season, it became necessary for the National Hurricane Center to use the Greek alphabet in naming storms when a tropical wave near the Windward Islands became a depression and intensified to a tropical storm south of Puerto Rico on October 22. Alpha moved northwest and made landfall in the Dominican Republic with 50 mph winds and heavy rains resulting in 20 deaths from flooding. Weakening to a tropical depression, Alpha moved north into the southern Bahamas and was absorbed by the circulation of Hurricane Wilma on October 24.

Hurricane Beta—Developed from a tropical wave over the extreme southwestern Caribbean Sea becoming a tropical storm on October 27 and a hurricane on October 29. After causing significant damage on Providencia Island, Beta turned to the west and southwest and strengthened to a category 3 hurricane on October

TABLE 9-5. WATCH AND WARNING SUMMARY FOR HURRICANE WILMA BEGINNING AT THE POINT THAT THE HURRICANE TURNED TOWARD THE SOUTHWEST FLORIDA COAST, OCTOBER 20-24.

Date	Time (GMT)	Action	Location
10/20	0300	Tropical Storm Warning issued	Cuba for the provinces of La Habana, Ciudad de la Habana, Pinar del Rio, and the Isle of Youth
10/22	1500	Hurricane Watch issued	All of the Florida Keys including the Dry Tortugas and Florida Bay
10/22	2100	Hurricane Watch issued	West coast of Florida from Longboat Key southward and the east coast of Florida from Titusville southward including Lake Okeechobee
10/22	2100	Hurricane Watch changed to Hurricane Warning	Cuban provinces of Ciudad de la Habana, La Habana, and Pinar del Rio
10/23	2100	Tropical Storm Watch issued	Florida from N of Longboat Key to the Steinhatchee River and from N of Titusville to Fernandina Beach
10/23	0300	Hurricane Watch changed to Hurricane Warning	The Florida Keys including the Dry Tortugas and Florida Bay, the West coast of Florida from Longboat Key southward, the East coast of Florida from Jupiter Inlet southward including Lake Okeechobee
10/23	0300	Hurricane Watch issued	NW Bahamas including the Abacos, Andros Island, Berry Islands, Bimini, Eleuthera, Grand Bahama Island, and New Providence
10/23	0900	Hurricane Watch changed to Hurricane Warning	Florida from North of Jupiter Inlet to Titusville
10/23	0900	Tropical Storm Watch changed to Tropical Storm Warning	Florida from North of Titusville to Flagler Beach
10/23	1200	Hurricane Warning issued	NW Bahamas including the Abacos, Andros Island, Berry Islands, Bimini, Eleuthera, Grand Bahama Island, and New Providence
10/24	0300	Tropical Storm Watch changed to tropical Storm Warning	Florida from North of Titusville to St. Augustine
10/24	1500	Tropical Storm Watch discontinued	Florida from North of St. Augustine to Fernandina Beach
10/24	1700	Tropical Storm Warning discontinued	Florida from North of Longboat Key to the Steinhatchee River
10/24	1830	Hurricane Warning changed to Tropical Storm Warning	The Florida Keys including the Dry Tortugas and Florida Bay, Florida West coast from Longboat Key southward and Florida East coast from South of Florida city including Lake Okeechobee
10/24	1830	Watches/Warnings discontinued	Cuba
10/24	2100	Warnings discontinued	Florida
10/24	2100	Warnings discontinued	Bahamas

Source: Richard J. Pasch, Eric S. Blake, Hugh D. Cobb III and David P. Roberts. National Hurricane Center, Miami, FL, 2006.

30. After weakening to a category 2 hurricane, Beta made landfall on the central east coast of Nicaragua.

Tropical Storm Gamma—Formed from a tropical wave which passed through the southern windward Islands on November 13 and then became tropical depression 27 on the 14th. It briefly approached tropical storm strength on November 15 before degenerating back into a tropical depression on November 17. Gamma finally formed a tropical storm on the 18th of November near the northern coast of Honduras and drifted north with 50 mph winds east of Roatan Island. Heavy rainfall with flash floods and mud slides occurred in Honduras and Belize resulting in 37 deaths and destruction of bridges.

Tropical Storm Delta—Formed from an extra-tropical low over the central Atlantic Ocean on November 19. It moved east and northeast to a point 800 miles southwest of the Azores Islands on November 22, then turned south and developed into a subtropical storm. Continuing south on November 23, Delta became a tropical storm with winds estimated at 70 mph. Delta moved erratically for several days and again became an extra-tropical low on November 28 and brought wind gusts of hurricane force to the Canary Islands before weakening and moving into Morocco on November 29. Seven deaths resulted in the Canary Islands from the extra-tropical phase of Delta.

Tropical Storm Epsilon—Also developed from an extra-tropical low over the central Atlantic midway between Bermuda and the Azores. At the end of November it was located several hundred miles east of Bermuda and moving east-northwest over open waters.

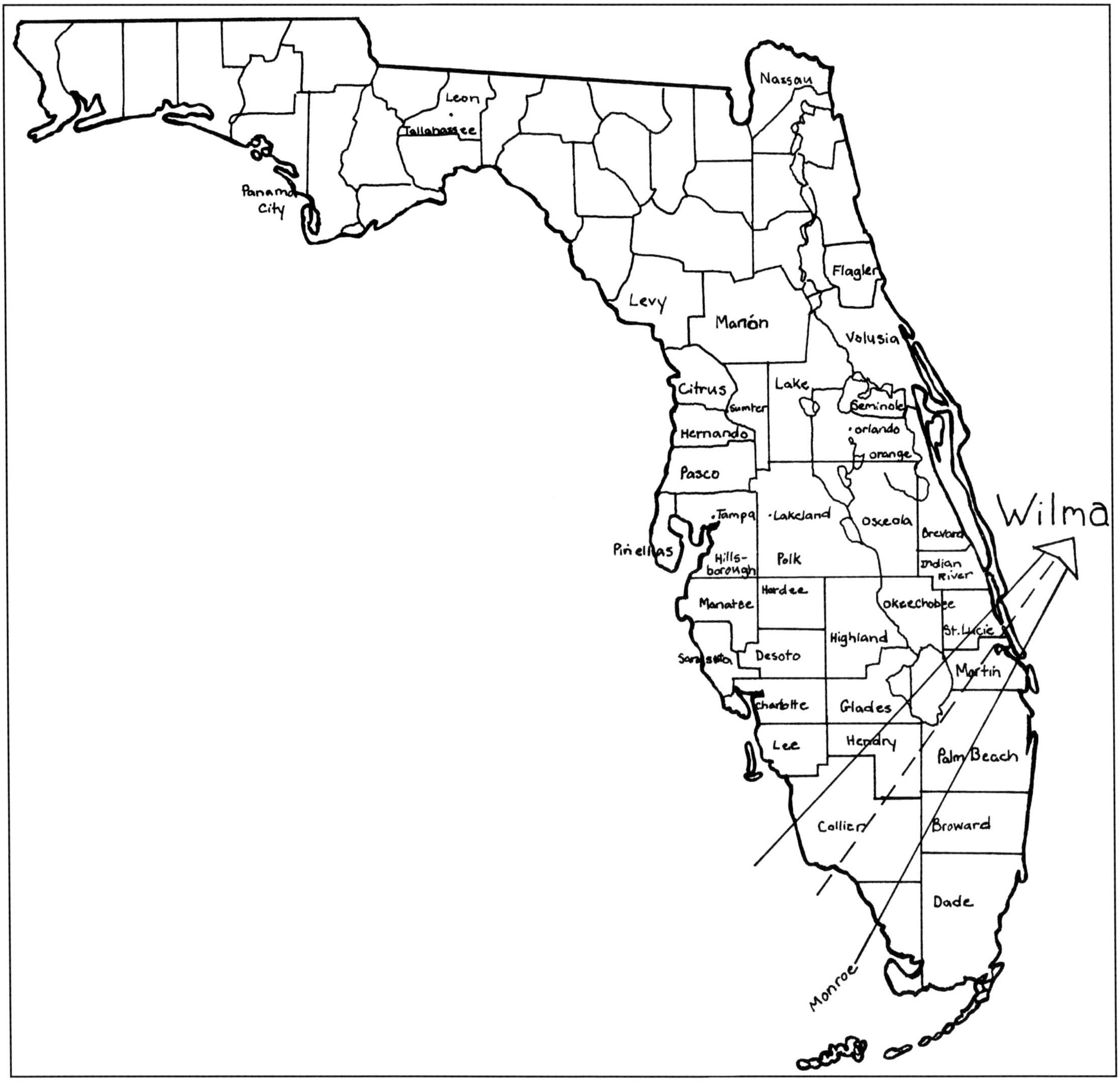

Figure 9-5. The path of Hurricane Wilma as she crossed south Florida, devastating groves in Collier, Hendry and Palm Beach counties, which had been largely spared by the 2004 hurricanes. Sketch by Brooke Elizabeth Attaway of Lakeland, FL.

CHAPTER 10

THE EFFECT OF HURRICANE WILMA ON THE ORANGE JUICE FUTURES MARKET

The 2004-05 Florida orange crop was 149.6 million 1-3/5 bushel boxes, drastically reduced from 242 million 1-3/5 bushel boxes, in 2003-04 as a result of Hurricanes Charley, Frances and Jeanne. The decrease in crop size resulted in an increase in the price of an orange juice contract from 55 cents per pound solids in June 2004 to 96 cents in June 2005 after briefly trading above $1.00 per pound solids in March 2005.

During the month of July 2005 orange juice traded in an 8¢ range from 98¢ to $1.06, after which there was a gradual decline in August to a close of 89.5¢ on August 31. During September the market increased steadily to close at $1.02 at the end of the month in anticipation of a manageable crop. This increase continued during October and when the National Agricultural Statistics Service (NASS) released an estimate of 190 million boxes on October 12, 2005, over 40 million boxes higher than the 2004-05 crop, but 52 million boxes lower than the 2003-04 crop, the juice price steadied and began to rise, further aided by the birth of Hurricane Wilma in the western Caribbean. When Wilma reached the category 5 level on the Saffir-Simpson Scale on October 19, 2005, juice traded as high as $1.145, and maintained this high level for the balance of the month as Wilma crossed through Collier, Hendry, Martin, St. Lucie and Palm Beach counties, the heart of the south Florida citrus industry. On October 28, 2005, the last trading day of the month, orange juice closed at over $1.167 per pound solids on a volume of 6,502 contracts.

In November, as damage reports poured in from the affected counties, the market held firm with open interest ranging from 32,344 to 36,458 contracts and daily volume from as low as 1,115 to as high as 4,660 contracts (Table 10-1). There was no official crop estimate on November 10, 2005, so the estimate continued to be 190 million boxes, although major losses were understood. These losses were confirmed when the orange crop estimate was reduced to 162 million boxes on December 9, 2005, and the futures market increased from $1.242 to $1.266 and rose an additional four cents to $1.306 on the following Monday.

Looking at the full month of December (Table 10-2) open interest ranged from 37,665 contracts on December 15 to 30,590 contracts December 30, the last trading day of the month. Daily volume hit a low of 2,008 con-

TABLE 10-1. OPEN INTEREST, VOLUME AND DAILY CLOSE FOR ORANGE JUICE FUTURES' CONTRACTS FOR NOVEMBER 2005.

Date	Open Interest[1]	Volume	Daily Close
11/1	32,344	4,521	$1.157
11/2	33,104	4,117	1.190
11/3	34,152	2,782	1.191
11/4	34,645	2,186	1.218
11/7	34,896	2,040	1.235
11/8	34,670	1,595	1.227
11/9	34,746	1,964	1.220
11/10	35,408	1,614	1.231
11/11	35,801	1,705	1.224
11/14	35,901	1,119	1.215
11/15	35,866	1,739	1.201
11/16	36,458	2,241	1.220
11/17	36,586	1,155	1.203
11/18	35,518	4,660	1.118
11/21	35,806	1,838	1.189
11/22	35,624	3,446	1.116
11/23	34,814	2,457	1.116
11/28	34,890	2,743	1.118
11/29	34,791	1,836	1.128
11/30	34,181	4,340	1.201

[1]The total number of futures contracts or option contracts that have not been exercised, expired or fulfilled by delivery.
Source: Citrus Associates, of the New York Board of Trade.

TABLE 10-2. OPEN INTEREST, VOLUME AND DAILY CLOSE OF ORANGE JUICE FUTURES' CONTRACTS FOR DECEMBER 2005.

Date	Open Interest[1]	Volume	Daily Close
12/1	34,714	6,044	$1.231
12/2	35,490	4,631	1.265
12/5	35,307	3,491	1.205
12/6	35,626	2,008	1.215
12/7	35,804	2,048	1.233
12/8	36,238	3,371	1.245
12/9	36,778	6,659	1.266
12/12	37,067	4,325	1.306
12/13	36,942	5,486	1.304
12/14	37,328	5,282	1.284
12/15	37,665	3,791	1.283
12/16	36,639	8,623	1.276
12/19	36,339	2,887	1.264
12/20	37,085	9,492	1.251
12/21	37,248	7,637	1.247
12/22	37,371	4,335	1.255
12/23	37,225	4,768	1.242
12/27	34,711	9,873	1.198
12/28	33,266	5,919	1.206
12/29	32,267	5,143	1.240
12/30	30,590	7,761	1.257

[1]The total number of futures contracts or option contracts that have not been exercised, expired or fulfilled by delivery.
Source: Citrus Associates, of the New York Board of Trade.

tracts on December 6 and peaked at 9,492 contracts on December 20. The volume on December 30 was 7,761 contracts. The high for the Daily Close was $1.306 on December 12, after the crop estimate had been released, but dropped to $1.198 on the 27th.

CHAPTER 11

SOME HISTORIC HURRICANES[1]

The U.S. Highway 17 Hurricanes— A Comparison

The October 1944 Hurricane

No two hurricanes are alike. They may follow a similar course, but are never exactly the same. They may have similar forward speeds, but they vary. The sustained winds and peak gusts may be nearly the same, but they also vary. And, the size of the eye and width of the overall storm, including tropical storm force winds will also differ. This is especially true of the three Highway 17 hurricanes. The October 1944 hurricane had an immense eye often over 40 miles in diameter with hurricane force winds stretching from the Atlantic to the Gulf of Mexico. Hurricane Donna was not as large as the 1944 hurricane, but was larger than Charley.

Again, it is difficult to compare the three hurricanes in terms of fruit loss. When Donna and the 1944 hurricane struck, Hendry, Lee, Charlotte, Collier and DeSoto counties were relatively minor citrus producing counties. However, when Charley stormed ashore in August 2004, DeSoto County had become one of Florida's most important counties in orange production, and Hardee County had doubled its 1960 production. While Charley's eye was relatively narrow, the damage to groves in Charlotte, DeSoto and Hardee counties was catastrophic to growers in its path.

The first of the Highway 17 hurricanes struck the state October 18-19, 1944. It had formed near Swan Island in the western Caribbean Sea on October 12 and passed over Grand Cayman Island on the 13th and 14th with wind gusts to 118 mph. Turning north the hurricane crossed Cuba with 140 mph sustained winds and gusts to 160 mph, then continued over the Dry Tortugas.

Shifting to the north-northeast the storm came ashore south of Sarasota with an eye estimated 40 miles in diameter.

Following U.S. 17 north the hurricane was preceded by several tornadoes as it passed through DeSoto, Hardee and Polk counties. Dangerous winds extended 200 miles to the east of the center and 100 miles to the west, encompassing the entire Florida peninsula. Wind gusts of 100 mph or more were felt in Tampa to the west and Orlando to the east. The heaviest wind damage occurred across a 30 mile wide strip.

Florida counties on the stronger east side of the storm included, Lee, Charlotte, Hendry, DeSoto, Hardee, Polk, Osceola, Lake, Orange, Seminole, Marion and Volusia.

High winds whipped groves unmercifully, with greatest damage to grapefruit and tall seedling orange trees. The greatest loss of citrus was in DeSoto, Hardee, Polk, Orange and Lake counties, with the total loss of all varieties predicted to be 23.7 million boxes out of a total estimated crop of 92.7 million boxes (Table 11-1).

A statement was issued by the Florida Citrus Commission indicating that Florida had lost approximately 50% of its grapefruit and from 20 to 25% of its oranges. The report further indicted that additional losses might result as weakened fruit dropped to the ground later.

Hurricane Donna—the Second of the Highway 17 Hurricanes

Hurricane Donna struck Florida September 10-11, 1960, sixteen years after the October 17-18, 1944 hurricane and 44 years before Charley, the third of the Highway 17 hurricanes. At the time it was considered by many to be the most destructive storm to impact the Florida citrus industry since the great east coast hurricane of August 1949. Its tremendous power was demonstrated by the fact that it maintained hurricane force winds from the time it crossed the southwest Florida coastline until it dissipated in New England. The following 44 years did not produce a hurricane capable of causing such widespread damage to the Florida citrus industry until Charley arrived in August 2004.

Donna was a classic Cape Verde hurricane. It formed from an easterly wave west of Dakar, Senegal and reached hurricane intensity on September 1, 1960. Moving west northwest, Donna passed through the Leeward Islands the evening of September 4, causing severe damage. Fortunately, Puerto Rico, the Virgin Islands, and the Turks Islands were south of the center and suffered little damage. However, Mayaguana, Acklins Island, Fortune Island and Ragged Island in the Bahamas suffered major damage to houses and buildings from 12 or more hours of hurricane force winds with gusts in excess of 150 mph.

When the eye crossed the Florida Keys after midnight on September 10, sustained winds were 140 mph with

[1]Attaway, J. A. 1999. Hurricanes and Florida Agriculture. Fla. Sci. Source, Lake Alfred, FL.

TABLE 11-1. CITRUS FRUIT LOSSES FROM THE HURRICANE OF OCTOBER 18-19, 1944 ACCORDING TO OFFICIAL GOVERNMENT FIGURES.

Fruit type	October estimate in boxes	Actual production in boxes	Loss in boxes	Percent loss
	- - - - - - - - - - - - -	millions	- - - - - - - - - - - - -	
Oranges	52.0	42.8	9.2	17.7
Grapefruit	36.0	22.3	13.7	38.1
Tangerines	4.7	3.9	0.8	17.0
Total	92.7	69.0	23.7	25.6

Source: Florida Agricultural Statistics Service, 1997.

gusts estimated at 180 mph. It began to recurve northward at a forward speed of 8 miles per hour and aimed its ferocious wind gusts toward Naples and Ft. Myers.

Track Across Florida

It would be difficult to plot a storm track with greater potential for damage to Florida citrus than that followed by Hurricane Donna. The eye passed over Naples and Fort Myers, then followed U.S. Highway 17 north over Arcadia, Wauchula, Bartow, Winter Haven, just west of Orlando and DeLand and reentered the Atlantic Ocean at Flagler Beach, with the powerful right quadrant passing through Collier, Lee, Charlotte, Hendry, Glades, DeSoto, Hardee, Highlands, Polk, Lake, Orange and Volusia counties. At Fort Myers, the sustained wind was 92 mph with gusts to 121 mph. At Everglades City and Naples, sustained winds were estimated at 100 mph with gusts from 140 to 150 mph (Table 11-2).

The Damage

Prior to the hurricane, there were unofficial predictions that Florida would see its first 100 million box orange crop and 40 million box grapefruit crop during the 1960-61 season (Florida Grower and Rancher, 1960). However, Donna's strong, gusty winds prevented this from occurring. The September 19, 1960 *Triangle* provided a complete, county-by-county estimate of hurricane damage similar to that provided after major freezes. The results were as follows, beginning with the most heavily damaged southwest Florida counties and ending with the least damaged north central counties:

Lee and Hendry Counties. Very severe damage to citrus occurred in this area. Ninety to 100% of the grapefruit, 60 to 65% of the early and midseason oranges and 50% of the Valencias were lost. There was a 15 to 20% tree loss and a very heavy foliage loss.

Collier County. No report due to scarcity of plantings in that county in 1960.

Charlotte County. Loss of grapefruit was set at 75% and oranges 40 to 50%. Tree and foliage damage in older groves was severe.

Glades County. No report due to few plantings in that county in 1960.

DeSoto County. Extensive damage to citrus occurred in this county. Damage to the orange crop was estimated at 40 to 50% with the loss on some individual

TABLE 11-2. METEOROLOGICAL DATA AT REPRESENTATIVE FLORIDA STATIONS AS HURRICANE DONNA MOVED FROM SOUTH TO NORTH THROUGH THE FLORIDA PENINSULA, SEPTEMBER 9-11, 1960.

Station	Minimum pressure (inches)	Maximum sustained winds (mph) directions	Time	Wind gusts (mph)
Tavernier[1]	28.80	120	—	160[3]
Everglades City	28.15	150 NE	9:15 A.M.	150[3]
Ft. Myers	28.08	92 NE	1:31 P.M.	121
Naples[2]	28.04	100[3]	12:20 P.M.	150[3]
Homestead	29.17	60 ESE	4:36 A.M.	89
Tampa	29.11	62 N	9:40 P.M.	75
Lakeland	28.60	68 NE	9:35 P.M.	90[3]
Leesburg	28.88	62 NE	1:30 A.M.	96
Orlando	28.66	46 ENE	2:36 A.M.	69
Daytona Beach	28.73	58 ENE	3:42 A.M.	86

[1]Sustained wind above 120 mph for 45 minutes, above 100 mph for 7 hours. Gusts above 100 mph for 13 hours.
[2]At Naples, the station was abandoned before time of highest winds. Eye over city 12-1 P.M.; 3 to 4 feet of water over most of city.
[3]Estimated.
Source: Climatological Data, September 1960.

groves even higher. The grapefruit loss was from 80 to 90% and the tangerine loss at 25 to 30%.

Hardee County. Damage was very severe. Fruit loss estimates were 75% for grapefruit, 40 to 50% for oranges and 20 to 25% for the tangerine crop.

Highlands County. Fruit loss was very heavy with grapefruit estimates ranging from 75 to 90%. Marsh grapefruit suffered a heavier loss than Duncans. Estimates of orange losses ranged from 15 to 35% and tangerines at 20%.

Sarasota County. Loss of grapefruit was estimated at 40 to 50% with some groves losing as much as 75% of the fruit. Loss of early and midseason oranges was 25% and Valencias 15%.

Polk County. Grapefruit losses ranged between 20 and 60%, averaging 35 and 40%. Early and midseason orange variety losses averaged 20%.

Hillsborough County. Orange loss was 15 to 20%, and grapefruit 35 to 40%. Tangerines 15 to 20%.

Broward County (Davie). Estimates were 20% loss of grapefruit, 15 to 20% of early and midseason oranges and 10% of Valencias.

St. Lucie and Indian River Counties. There was about 5% grapefruit loss, but no orange loss and no tree or foliage damage.

Brevard County. Grapefruit loss of about 20% and an orange loss of 3 to 5%.

Pasco County. About 10 to 15% of the grapefruit and 2 to 3% of the oranges were on the ground. Early and midseason oranges, especially seedlings, and Marsh grapefruit were hardest hit. Individual groves lost as much as 50% of their fruit and as little as 1%.

Lake County. Damage was light. Grapefruit loss was put at about 10%, oranges 1 to 2% and losses to tangerines, tangelos and Temples was negligible.

Orange and Seminole Counties. Grapefruit loss was estimated to be 10 to 20%, orange loss 3 to 5%, and tangerine loss 5 to 8%. Osceola County experienced somewhat heavier losses. Grapefruit was set at 50%, oranges at 10 to 15%, and tangerines 8 to 10%.

Hernando and Pinellas Counties. Orange loss in these counties was estimated at 1 to 2% and grapefruit loss at 5 to 7%.

Citrus and Sumter Counties. These counties experienced only gale winds and hardly any damage to either fruit or trees.

Marion County. There was only a small amount of fruit and tree damage.

Volusia County. Damage to citrus in the western part of the county was not severe. Less than 3% of the fruit was lost and tree damage was negligible. However, in east Volusia County, grapefruit loss was estimated at 20% and oranges 5 to 10%.

Noting from the widespread damage across many counties from east to west, Donna was a larger hurricane than Charley and therefore more destructive. Charley was very intense, and the area within a few miles of the eye was devastated, but as the hurricane moved through the central Florida counties, groves a few miles from the center escaped serious damage.

The Great Indian River Hurricane of 1949—A Comparison

The August 26-27, 1949 Hurricane

The August 1949 hurricane was one of the most destructive of the modern hurricanes to impact peninsular Florida, particularly for the havoc it wreaked on the Florida citrus industry. Striking first at the bountiful Indian River grapefruit producing counties with the full force of its 150 mph winds, the storm then ripped through Polk and Highlands counties with wind gusts still approaching 100 mph. Later it passed through Lake and Orange counties with wind gusts still at minimal hurricane force.

The Meteorology

The 1949 hurricane formed from an easterly wave 125 miles northeast of St. Martin in the Leeward Islands (Norton, 1949), and moved west north-west reaching the Bahamas on August 25, at which point a well-developed center had formed. Pointing its fury toward the southeast Florida coast, the eye was felt first at Delray Beach, then moved over the West Palm Beach airport, where the anemometer was blown down by sustained winds reaching 110 mph with gusts to 125 mph. A private observer at Palm Beach recorded gusts up to 155 mph. Sustained winds of 125 mph were recorded at Stuart and 83 mph at Cape Canaveral.

After leaving thousands of boxes of grapefruit on the ground, the hurricane took direct aim at Central Florida. Moving away from the warm waters needed to fuel its engine, the hurricane diminished in intensity as it moved over the ridge with 90 mph sustained winds at Lake Placid, 100 mph gusts at Bartow and 75 mph at Clermont.

The Damage

Polk, Indian River, St. Lucie and Highlands counties suffered the greatest damage, with grapefruit losses at 90% in St. Lucie County, 75% in Indian River County and 55 to 65% in Highlands and Polk counties; and, orange losses at 40% in St. Lucie County and 30% in

Indian River County, noting, however, that these two counties combined produced only 5% of the early and midseason oranges and 8% of the Valencia oranges.

For groves in the direct path of the storm in Polk and Highlands counties, sources predicted the loss of 65 to 70% of Marsh grapefruit, 40 to 50% of Duncan grapefruit, 25% of early oranges and 10 to 15% of Valencia oranges, with a 15 to 17 million box loss of all varieties statewide, most of which would be grapefruit. Robert C. Evans, general manager of the Florida Citrus Commission, estimated the loss for the state at large at 13 to 17 million boxes, with 10 to 12 million boxes of grapefruit accounting for the bulk of the loss.

Final Reports

Grapefruit production in St. Lucie County dropped from 1,178,000 boxes for the 1948-49 season to only 359,000 boxes for 1949-50, the hurricane year, and then recovered to 1,239,000 boxes in 1950-51. Precipitous losses in grapefruit production also occurred in Indian River, Polk and Highlands counties with strong recoveries in the following season. Orange losses were not as severe except in St. Lucie County.

APPENDIX ONE

GLOSSARY OF TERMS

Anemometer—An instrument for measuring wind speed.

Atmospheric Pressure—The force exerted by a column of air over a specific area. It is usually measured by a barometer and expressed as inches of mercury or millibars (mb). The standard pressure at sea level is 29.92 inches of mercury or 1013.35 mb.

Barometer—An instrument for measuring atmospheric pressure.

Beaufort Wind Scale—A scale used historically by seamen to describe or estimate wind speed. Dunn and Miller (1964).

Cape Verde Hurricane—An Atlantic tropical cyclone that develops into a tropical storm near the Cape Verde Islands and becomes a hurricane before reaching the Caribbean.

Cyclone—A low pressure system whose winds blow counterclockwise in the northern hemisphere.

Easterly Wave—A low pressure trough moving in a westerly direction toward the Caribbean Islands or the southern coast of the United States. It may also be called a tropical wave.

Eye—The relatively calm center of a hurricane. The average diameter of a hurricane eye is 14 miles. However, hurricane eyes have been known to range from only 4 miles in a developing hurricane to 20-25 miles in a large, mature hurricane (Dunn and Miller, 1960).

Gale—A wind equal to or exceeding 39 mph.

Hurricane—A tropical cyclone with maximum sustained winds of 74 mph or greater.

Hurricane Force Winds—Winds of 74 mph or greater.

Hurricane Warning—A warning that hurricane force winds may be felt in a specified area within 24 hours.

Hurricane Watch—An indication that hurricane force winds may be felt in a specified area within 36 hours.

Intertropical Convergence Zone—Zone near the equator along which the opposing trade winds of the northern and southern hemispheres meet.

Maximum Sustained Winds—Maximum wind speed sustained over a one minute period.

Millibar (mb)—A unit of pressure.

Peak Wind Gust—Highest wind speed. According to Dunn and Miller (1964), wind speeds in hurricanes fluctuate so rapidly that peak gusts in hurricanes are difficult to record accurately. Peak wind gusts may exceed the highest sustained wind speed by 30 to 50%. Therefore, gusts of over 200 mph could occur in hurricanes with maximum sustained winds of 150 mph. Such gusts are responsible for the massive destruction in severe hurricanes.

Recurvature—The change in direction of an Atlantic Basin hurricane or tropical storm from a west or northwest course to a north or northeasterly course.

Quadrant—A quarter-circle (90%) of the wind field of a hurricane. In an Atlantic Basin hurricane, the strongest winds are found in the northeast quadrant.

Tropical Cyclone—A cyclone which has developed over tropical waters.

Tropical Depression—A tropical cyclone with winds up to 38 mph.

Tropical Disturbance—An area of thunderstorms over tropical waters. May move as part of a tropical wave.

Tropical Storm—A tropical cyclone with maximum sustained winds ranging from 39-73 mph (34-63 knots).

Tropical Storm Warning—A warning that tropical force winds may be felt in a specified area within 24 hours.

Tropical Storm Watch—An warning that tropical force winds may be felt in a specified area within 36 hours.

Tropical Wave—A low pressure trough moving from east to west. Also called an easterly wave. In the Atlantic Basin, the wave will usually be moving toward the Carribean Islands.

Saffir-Simpson Scale—A scale for measuring the relative destructive potential of hurricanes. See Appendix Two.

UTC—Stands for Universal Time Coordinated, formerly called Greenwich Mean Time (GMT) and Zulu Time (Z). This is the time at the Prime Meridian (0° Longtitude) given in hours and minutes on a 24 hour clock.

Wind Shear—Winds aloft which may shear the top of a developing storm thus preventing hurricane formation.

APPENDIX TWO

THE SAFFIR-SIMPSON SCALE

The Saffir-Simpson Scale is used by the National Hurricane Center to characterize hurricanes by intensity. The scale ranges from 1 to 5, with 1 being a minimal hurricane and 5 being the most violent and destructive hurricanes such as Hurricanes Katrina and Rita in 2005, Hurricane Andrew in 1992, and the Florida Keys hurricane in 1935. The Scale is shown in the following table with approximate damage which might be expected to citrus trees.

SAFFIR-SIMPSON SCALE TABLE

Scale Number	Wind Speed (mph)	Predicted Damage
1	74-95	Some loss of leaves and fruit, particularly in exposed areas.
2	96-110	Considerable loss of leaves and fruit with foliage damage in exposed areas.
3	111-130	Heavy loss of foliage and fruit with many trees uprooted.
4	131-155	Damage major, trees stripped of all foliage and fruit, many trees uprooted.
5	Over 155	Damage extremely severe with groves completely destroyed, with trees blown away from property.*

*An example was lime trees in south Dade County after Hurricane Andrew.
Source: National Hurricane Center, Miami, FL.

APPENDIX THREE

NAMES FOR FUTURE HURRICANES AND TROPICAL STORMS

The following names will be used in the five year period beginning in 2006. Note that short distinctive names are preferred to avoid confusion. Storms prior to 1953 were not named. Initially only women's names were used, but since 1979 both men's and women's names are used.

PROPOSED HURRICANE NAMES FOR THE YEARS 2006-2010

2006	2007	2008	2009	2010
Alberto	Andrea	Arthur	Ana	Alex
Beryl	Barry	Bertha	Bill	Bonnie
Chris	Chantal	Christobal	Claudette	Colin
Debby	Dean	Dolly	Danny	Danielle
Ernesto	Erin	Edouard	Erika	Earl
Florence	Felix	Fay	Fred	Fiona
Gordon	Gabrielle	Gustav	Grace	Gaston
Helene	Humberto	Hanna	Henri	Hermine
Isaac	Ingrid	Ike	Ida	Igor
Joyce	Jerry	Josephine	Joaquin	Julia
Kirk	Karen	Kyle	Kate	Karl
Leslie	Lorenzo	Laura	Larry	Lisa
Michael	Melissa	Marco	Mindy	Matthew
Nadine	Noel	Nana	Nicholas	Nicole
Oscar	Olga	Omar	Odette	Otto
Patty	Pablo	Paloma	Peter	Paula
Rafael	Rebekah	Rene	Rose	Richard
Sandy	Sebastien	Sally	Sam	Shary
Tony	Tanya	Teddy	Teresa	Tomas
Valerie	Van	Vicky	Victor	Virginie
William	Wendy	Wilfred	Wanda	Walter

Source: National Hurricane Center, Miami, FL.

APPENDIX FOUR

THE 30 MOST DEADLY HURRICANES IN RECENT U.S. HISTORY[1]

The greatest loss of life from Atlantic hurricanes striking the U.S. Mainland occurred during the following hurricanes. The category is determined by the strength of the hurricane at landfall.

DEADLY HURRICANES SINCE 1900

Rank/Name	Year	Category	Deaths
1. Texas (Galveston)	1900	4	8,000
2. Southeast Florida	1928	4	1,836
3. Florida Keys & Texas	1919	4	600
4. New England	1938	3	600
5. Florida Keys	1935	5	408
6. Audrey	1957	4	390
7. Northeast United States	1944	3	390
8. Louisiana (Grande Isle)	1909	4	350
9. Louisiana (New Orleans)	1915	4	275
10. Texas (Galveston)	1915	4	275
11. Camille	1969	5	256
12. Florida (Miami, Pensacola)	1926	4	243
13. Diane	1955	1	184
14. Southeast Florida	1906	2	164
15. Pensacola, Ala., Miss.	1906	3	143
16. Agnes (Florida, NE U.S.)	1972	1	122
17. Hazel	1954	4	95
18. Betsy (S.E. Florida, LA)	1965	3	75
19. Carol	1954	3	60
20. Floyd	1999	2	56
21. Isabel	2003	2	55
22. Ivan	2004	3	52
23. S.E. Fla., La., Miss.	1947	4	51
24. Donna (Fla., Eastern U.S.)	1960	4	50
25. Ga., S.C., N.C.	1940	2	50
26. Carla	1961	4	46
27. Texas (Velasco)	1909	3	41
28. Texas (Freeport)	1932	4	40
29. South Texas	1933	3	40
30. Frances	2004	2	38

[1]At the time of publication, final death totals for 2005 hurricanes were not available.
Source: National Hurricane Center, Miami, FL.

APPENDIX FIVE

RETIRING OF HURRICANE NAMES

The list of names for tropical storms and hurricanes is established by an international committee of the World Meteorological Organization, with the Atlantic hurricane list rolled over every seventh year. However, in the case of extremely severe hurricanes, a name will be dropped from the list and no longer used in the future.

The following are names of Atlantic Ocean, Carribean Sea and Gulf of Mexico hurricanes which have been retired.

RETIRED HURRICANE NAMES

Year	Retired Names(s)	Year	Retired Name(s)
1954	Carol[1], Hazel	1980	Allen
1955	Connie, Diane, Ione, Janet	1983	Alicia
1957	Audrey	1985	Elena, Gloria
1960	Donna	1988	Gilbert, Joan
1961	Carla, Hattie	1989	Hugo
1963	Flora	1990	Diana, Klaus
1964	Cleo, Dora, Hilda	1991	Bob
1965	Betsy	1992	Andrew
1966	Inez	1995	Luis, Marilyn, Opal, Roxanne
1967	Beulah	1996	Ceasar, Fran, Hortense
1968	Edna	1998	Georges, Mitch
1969	Camille	1999	Floyd, Lenny
1970	Celia	2000	Keith
1972	Agnes	2001	Allison, Iris, Michelle
1974	Carmen, Fifi	2002	Isidore, Lili
1975	Eloise	2003	Fabian, Isabel, Juan
1977	Anita	2004	Charley, Frances, Ivan, Jeanne
1979	David, Frederic	2005	Dennis, Katrina, Rita, Stan, Wilma

[1]The name "Carol" was used again to denote a hurricane in the mid-Atlantic Ocean in 1965. However, because the name does not appear after that time, it is assumed that the name was retired retrospectively for the damages caused by the 1954 storm of the same name.
Source: National Hurricane Center, Miami, FL.

APPENDIX SIX

THE MOST INTENSE U.S. HURRICANES 1900-2004

In terms of minimum pressure, the following are the most intense hurricanes to strike the United States mainland since 1900.

MOST INTENSE U.S. HURRICANES, 1900-2004

Rank/Name[1]	Year	Category	Minimum Pressure (in millibars)	Minimum Pressure (in inches)
1. Florida Keys	1935	5	892	26.35
2. Camille	1969	5	909	26.84
3. Andrew	1992	5	922	27.23
4. Florida Keys, Texas	1919	4	927	27.37
5. Lake Okeechobee	1928	4	929	27.43
6. Donna	1960	4	930	27.46
7. Texas (Galveston)	1900	4	931	27.49
7. Louisiana (Grande Isle)	1909	4	931	27.49
7. Louisiana (New Orleans)	1915	4	931	27.49
7. Carla	1961	4	931	27.49
12. Hugo	1989	4	934	27.58
14. Frances	2004	4	935	27.61
14. Florida (Miami, Pensacola,	1926	4	935	27.61
Miss., Ala.)	2002	4	938	27.70
16. Lili	1954	4	938	27.70
16. Hazel (SC, NC)	1954	5	940	27.76
18. Southeast Fla., La., Miss.	1947	4	941	27.79
20. Charley	2004	4	941	27.79
20. Texas	1932	3	942	27.82
22. Gloria (Eastern US)	1985	3	942	27.82
22. Opal (NW FL, AL)	1995	4	945	27.91
24. Audrey (SW LA, N TX)	1957	4	945	27.91
24. Texas (Galveston)	1915	3	945	27.91
24. Celia (S TX)	1970	3	945	27.91
24. Allen (S TX)	1980	3	946	27.94
28. New England	1938	3	946	27.94
28. Frederic (AL, MS)	1979	3	947	27.97
30. Ivan	2004	3	947	27.97

[1]Hurricanes were not named before 1950.
Source: National Hurricane Center, Miami, FL.

APPENDIX SEVEN

THE COSTLIEST U.S. HURRICANES 1900-2004

The 30 costliest tropical cyclones to strike the U.S. mainland, with addendum for Hawaii, Puerto Rico, and the U.S. Virgin Islands. Damages are listed in U.S. dollars and are not adjusted for inflation.

Rank	Hurricane	Year	Category	Damage
1	Andrew (SE FL, SE AL)	1992	5	$25,500,000,000
2	Charley (SW FL)	2004	4	15,000,000,000
3	Ivan (AL/NW FL)	2004	3	14,200,000,000
4	Frances (FL)	2004	2	8,900,000,000
5	Hugo (SC)	1989	4	7,000,000,000
6	Jeanne (FL)	2004	3	6,900,000,000
7	Allison (N TX)	2001	TS[a]	5,000,000,000
8	Floyd (Mid-Atlantic & NE U.S.)	1999	2	4,500,000,000
9	Isabel (Mid-Atlantic)	2003	2	3,370,000,000
10	Fran (NC)	1996	3	3,200,000,000
11	Opal (NW FL< AL)	1995	3	3,000,000,000
12	Frederic (AL, MS)	1979	3	2,300,000,000
13	Agnes (FL, NE U.S.)	1972	1	2,100,000,000
14	Alicia (N TX)	1983	3	2,000,000,000
15	Bob (NC, NE U.S.)	1991	2	1,500,000,000
16	Juan (LA)	1985	1	1,500,000,000
17	Camille (MS, SE LA, VA)	1969	5	1,420,700,000
18	Betsy (SE FL, SE LA)	1965	3	1,420,500,000
19	Elena (MS, AL, NW FL)	1985	3	1,250,000,000
20	Georges (FL Keys, MS, AL)	1998	2	1,155,000,000
21	Gloria (Eastern US)	1985	3	900,000,000
22	Lili (SC, AL)	2002	1	860,000,000
23	Diane (NE U.S.)	1955	1	831,700,000
24	Bonnie (NC, VA)	1998	2	720,000,000
25	Erin (NW FL)	1995	2	700,000,000
26	Allison (N TX)	1989	TS[a]	500,000,000
27	Alberto (NW FL, GA, AL)	1994	TS[a]	500,000,000
28	Frances (TX)	1998	TS[a]	500,000,000
29	Eloise (NW FL)	1975	3	490,000,000
30	Carol (NE U.S.)	1954	3	461,000,000
Addendum (Rank is independent of other events in group)				
15	Georges (USVI, PR)	1998	3	1,800,000,000
15	Iniki (Kaua'i, HI)	1992	Unknown	1,800,000,000
17	Marilyn (USVI, PR)	1995	2	1,500,000,000
21	Hugh (USVI, PR)	1989	4	1,000,000,000
26	Hortense (PR)	1996	1	500,000,000

Source: National Hurricane Center, Miami, FL.

Appendix Eight

A BRIEF SUMMARY OF THE SECTION 32 FEDERAL DISASTER PROGRAM: FLORIDA CITRUS PROGRAM

Eligible Citrus

The Florida Citrus Program provides assistance to eligible producers that suffered citrus losses and tree damage as a result of hurricanes Charley, France, and/or Jeanne. Assistance is based on the level of loss for each grove.

Location Tiers

Eligible groves are those located with one of four tiers established geographically based upon the level of destruction. Each band was established at the national level and is being provided to each County FSA Office. The delineations on the map will be used to assist producers during the sign-up period.

Certified Loss Level

Each applicant must certify on the citrus application (FSA-439) the actual level of loss experienced for the grove because of hurricanes Charley, Frances, and/or Jeanne. The certification of loss will reflect the actual loss of production and tree damage and cannot exceed the loss for the band in which the grove is physically located unless the local FSA County Committee (COC) approves the loss according to COC Adjustments.*

For example, a grove physically located in Band 1 (area with the greatest destruction) experienced losses of production and associated tree damage of 60 percent. The producer will certify acres in Tier 2 even though the grove is physically located in Band 1. See COC Adjustments* for grove losses that are greater than the tier and band for that grove.

Payment Calculations

The chart below lists the payment rates and percentages of the payment that are subject and not subject to payment limitations for each tier. The rate of uninsured producers of those without NAP coverage is 95 percent of the payment rate for insured producers and those with NAP coverage. The portion of payment associated with tree damage and cleanup is not subject to the $80,000 per person payment limit.

*COC Adjustments

Bands of destruction have been determined by DAFP. However, it is recognized that the hurricanes spawned tornadoes and other unusual occurrences outside those bands. COC may accept any documentation available which to their satisfaction substantiates the producer's claim of a higher level of loss than the groves band and associated tier. This documentation may include, but not limited to, crop insurance appraisals, photos, and verification through field visits, news articles, production records, sales records, or COC knowledge of the affected areas.

The approval and the documentation used for the approval will be documented in the COC minutes. A copy of the documentation will be attached to the application.

For more information, visit ftp://ftp.fsa.usda.gov/public/notices/DAP_203.pdf.

Tier 1-4	Payment Rate Per Acre for Participants with Insurance or NAP Coverage	Payment Rate Per Acre for Participants without Insurance or NAP Coverage	Production Loss: (% of Payment Subject to $80,000 Limitation (SL))	Tree Damage and Cleanup: (% of Payment not Subject to $80,000 Limitation (NL))
Tier 1: 75% or more crop loss and associated tree damage	$1,500	$1,425	55%	45%
Tier 2: 50-74% crop loss and associated tree damage	$1,000	$950	60%	40%
Tier 3: 35-74% crop loss and associated tree damage	$600	$570	64%	36%
Tier 4: 15% or greater tree loss and damage	$100	$95	0%	100%

REFERENCES

Adams, Jeff. 24 Aug. 2005. Personal Communication. Arcadia, FL.

Arnold, Dr. Calvin E. 2006. Personal Communication. U.S. Horticultural Research Laboratory, Ft. Pierce, FL.

Attaway, J. A. 1997. *A History of Florida Citrus Freezes*. Fla. Sci. Source, Lake Alfred, FL.

Attaway, J. A. 1999. *Hurricanes and Florida Agriculture*. Fla. Sci. Source, Lake Alfred, FL.

Becker, F.F. (Fran). 23 Aug. 2005. Personal Communication. Arcadia, FL.

Beven, John L. II. *Tropical Cyclone Report, Hurricanes Frances*. National Hurricane Center, Miami, FL. 17 Dec. 2004.

Bonnet, Chris. 23 Aug. 2005. Personal Communication. DeSoto County, FL.

Bouffard, Kevin. *Charley Broke the Rules for Citrus*. The Ledger, Lakeland, FL. 20 Aug. 2004.

Bouffard, Kevin. *Jeanne Squeezes Life Out of Citrus*. The Ledger. Lakeland, FL. 21 Aug. 2004

Broadaway, Dennis. 8 Sept. 2004. Newsletter to Haines City Citrus Growers Association Membership.

Bronson, Charles H. Letter to Governor Bush Seeking Declaration of Agricultural Disaster. 1 Sept. 2004. Triangle. 54:48, 2 Sept. 2004.

Callaham, Steve. 14 Sept. 2005. Personal Communication. Dundee, FL.

Caruso, Steve. 7 Sept. 2005. Personal Communication. Lake Wales, FL

Chadwick, Russ, Patricia Miller and David Helms. 2004. Hurricane Jeanne Landfall Observed by Citizen Weather Observer Program Stations.

Citrus Administrative Committee. 2006. Lakeland, FL.

Citrus Associates, New York Board of Trade, 2005.

Citrus Associates, New York Cotton Exchange, 2004.

Citrus Summary 2000-01.

Citrus Summary 2001-02.

Citrus Summary 2002-03.

Citrus Summary 2003-04.

Citrus Industry Magazine. 1960.

Climatological Data, Florida Section, 1928-1979.

Climatological Data, September 1960.

Commercial Citrus Inventory, Florida Agricultural Statistics Service

Counter, Charles. 2006. Personal Communication. Haines City, FL.

Crop Disaster Assistance Program.

Cusson, Jeff. 23 Jan 2006. Becker Holding Company. Ft. Pierce, FL.

Davis, Keith. 25 Aug. 2005. Personal Communiction. Wauchula, FL.

Dunn, G. E. and B. I. Miller. 1964. *Atlantic Hurricanes*. Louisiana State University Press, Baton Rouge, LA.

Farm Service Agency (FSA). U.S. Department of Agriculture. 13 Aug. 2004.

Florida Agricultural Statistics Service. 1997.

Florida Agricultural Statistics Service, Citrus Summary 2000-01.

Florida Agricultural Statistics Service, Citrus Summary 2001-02.

Florida Agricultural Statistics Service, Citrus Summary 2002-03.

Florida Agricultural Statistics Service, Citrus Summary 2003-04.

Florida Agricultural Statistics Service, Citrus Summary 2004-05.

Florida Citrus Commission.

Florida Citrus Mutual. Post-Hurricane Charley Grower Information Meetings. Triangle. 54:47, 20 Aug. 2004.

Florida Citrus Mutual. Mutual Applauds President Bush's Relief Plan. Triangle. 54:48, 2 Sept. 2004.

Florida Citrus Mutual. Estimated Fruit Losses from Hurricane Frances. 10 Sept. 2004.

Florida Field Office, National Agricultural Statistics Service, Orlando, FL.

Florida Grower and Rancher. 1960.

Florida State Newsletter, Farm Service Agency 2006.

Futch, Dr. Stephen H. 2 Sept. 2005. Personal Communication. Lake Alfred, FL.

Griffiths, Dr James T.. 2004. Citrus Grower Associates Newsletter. 4(24), 25 Aug. 2004.

Gunter, Dr. Dan. 21 Aug. 2005. The Continuing Impact of Hurricane Charley.

Hamner, George, Jr. 23 Jan. 2005. Personal Communication. Vero Beach, FL.

Hendrix, Charles M. (Charlie) Personal Communication. 14 Feb. 2006.

Hudson, Larry. 23 Aug. 2005. Personal Communication. Charlotte County, FL.

Hurley, R. Scott. 23 Jan. 2006. Personal Communication. Becker Holding Company. Ft. Pierce, FL.

Jackson, John. 15 Feb. 2006. Personal Communication. Tavares, FL.

Lakewood Community Association Newsletter. 8 Sept. 2004.

LaVigne, Andy. Letter to Florida Citrus Mutual Membership. 20 Aug. 2004.

LaVigne, Andy. 2006. Florida Citrus Mutual. Lakeland, FL.

Lawrence, Miles B. and Hugh D. Cobb. *Tropical Cyclone Report, Hurricane Jeanne*. National Hurricane Center. Miami, FL. 22 Nov. 2004, Revised 1 Jan. 2005.

Louis Dreyfus Citrus, Inc.

Manual, Mark. 2006. Personal Communication. Hardee County, FL.

Murphy, Michael S. 2006. Cooperative Producers, Inc. Immokalee, FL.

National Agricultural Statistics Service, Orlando, FL.

National Agricultural Statistics Service, Florida Field Office. 16 Sept. 2005.

Nelson, Gregory P. 17 Jan. 2006. Personal Communication. Ft. Pierce, FL.

Nelson, Gregory P. 2006. DNE World Fruit Sales, Ft. Pierce, FL.

Noninsured Crop Disaster Assistance Program (NAP).

Norton, 1949.

Oswalt, Chris. 8 Dec. 2005. Citrus Notes. Polk County, FL.

Pasch, Richard J., Daniel P. Brown and Eric S. Blake. 2005. *Tropical Cyclone Report, Hurricane Charley*. National Hurricane Center. 18 Oct. 2004, Revised 5 Jan. 2005.

Pasch, Richard. J., Eric S. Blake, Hugh D. Cobb, III and David P. Roberts. *Tropical cyclone Report, Hurricane Wilma*. National Hurricane Center. 12 Jan. 2006.

Post Storm Data Acquisition Missions. National Weather Service. 15-16 Aug. 2004 and 29 Aug. 2004.

Post-Hurricane Update—Selected Counties (NASS). 15 Sept. 2005.
Precautionary Boil Water Notice. 2004. City of Winter Haven, Sept. 7, 2004, Sept. 11, 2004.
Produce News. 17 Aug. 2004.
Roberts, Nat. 2006. Callery Judge Grove. Palm Beach County, FL.
Rouse, Dr. Bob. 2006. Southwest Florida Research and Education Center. Immokalee, FL.
Saffir-Simpson Scale.
Salisbury, Susan. 2005. Palm Beach Post. West Palm Beach, FL.
Scully, Dr. Brian. 23 Jan. 2006. Personal Communication. Indian River Research and Education Center, Ft. Pierce, FL.
Selph, James. 2005. Personal Communiction. Arcadia, FL.
South Region News. National Weather Service. 2005.
Spyke, Peter. 5-21 Sept. 2004. E-mails to Friends, Family and Clients during Hurricane Frances. Arapaho Citrus Management. Ft. Pierce, FL.
Spyke, Peter. 25 Sept. - 20 Oct. 2004. E-mails to Friends, Family and Clients during Hurricane Jeanne. Arapaho Citrus Management. Ft. Pierce, FL.
Spyke, Peter. 5 Sept. 2005. The Epilogue. A final E-mail to Friends, Family and Clients a year after Hurricane Jeanne. Arapaho Citrus Management. Ft. Pierce, FL.
Stewart, Stacy R. *Tropical Cyclone Report, Hurricane Ivan.* National Hurricane Center. 16 Dec. 2004, Updated 27 May 2005.
Sullivan, Matt. 23 Aug. 2005. Personal Communication. Charlotte County, FL.
Supplemental Appropriation's Bill.
Terry, Robert R. Florida Agricultural Statistics Service. 2004.
Townsend, Chet. 12-18 Aug. 2004. UltimateCitrus.com Citrus News. Ft. Denaud, FL. http://www.ultimatecitrus.com
Townsend, Chet. 6 Sept. 2004. UltimateCitrus.com Citrus News. Ft. Denaud, FL. http://www.ultimatecitrus.com
Townsend, Chet. 7 Sept. 2004. UltimateCitrus.com Citrus News. Ft. Denaud, FL. http://www.ultimatecitrus.com
Townsend, Chet. 2005. UltimateCitrus.com Citrus News. Fort Denaud, FL. http://www.ultimatecitrus.com
Triangle. 19 Sept. 1960.
U.S.D.A. Secretarial Disaster Declaration.
U.S.D.A. Corp and Livestock Reporting Service.
Unidentified grower.
www.agwise.net

INDEX